TOO CLOSE TO HOME?

..

McKIBBEN A. JACKINSKY

Too Close to Home?
Living with "drill, baby" on Alaska's Kenai Peninsula

The front- and back-cover photographs of Mount Iliamna
from two vantage points along the Sterling Highway
were taken by the author in October 2015.

A November 2015 map of oil and gas activity on the
west coast of the Kenai Peninsula can be viewed at:
http://dog.dnr.alaska.gov/GIS/Data/ActivityMaps/
CookInlet/CookInletOilAndGasActivityMap-201511.pdf.
The map is updated periodically—a basic link is:
http://dog.dnr.alaska.gov/GIS/ActivityMaps.htm.

Printed in the United States of America.

First printed May 2016.

Hardscratch Press, 658 Francisco Court,
Walnut Creek, California 94598-2231.

www.hardscratchpress.com

Library of Congress Control Number: 2016936635

ISBN: 978-0-9838628-6-4

2 4 6 8 9 7 5 3 1

DEDICATION

For my family – past, present, future
And for the land that is our home

TOO CLOSE TO HOME?

. .

Living with "drill, baby"
on Alaska's Kenai Peninsula

McKIBBEN A. JACKINSKY

A HARDSCRATCH PRESS BOOK

2016

CONTENTS

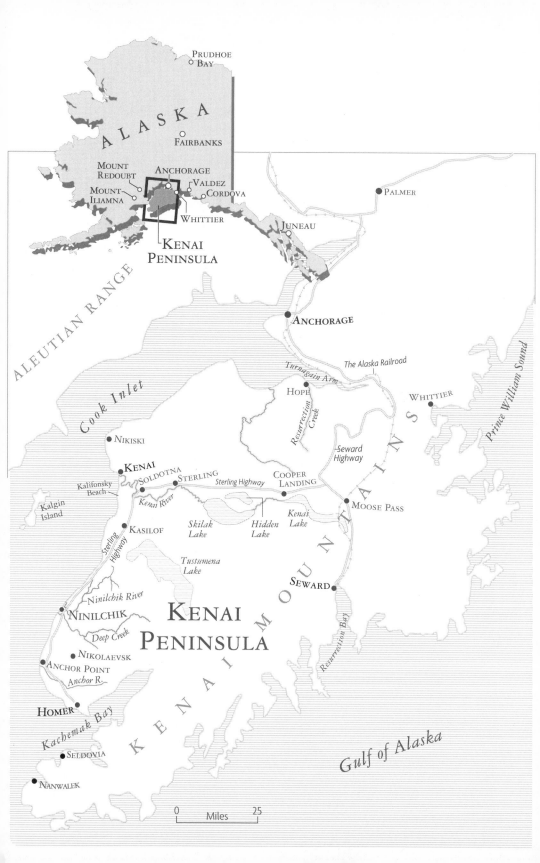

PRUDHOE
BAY

A L A S K A

FAIRBANKS

MOUNT
REDOUBT

ANCHORAGE
VALDEZ
CORDOVA

MOUNT
ILIAMNA

WHITTIER

KENAI
PENINSULA

JUNEAU

PALMER

ALEUTIAN RANGE

ANCHORAGE

Turnagain Arm

The Alaska Railroad

HOPE

Cook Inlet

Resurrection Creek

WHITTIER

Prince William Sound

Seward
Highway

NIKISKI

KENAI

SOLDOTNA

STERLING

Sterling Highway

COOPER
LANDING

Kalifonsky
Beach

Kenai River

Kenai Lake

MOOSE PASS

Kalgin
Island

KASILOF

Skilak
Lake

Hidden
Lake

Sterling Highway

Tustumena
Lake

K E N A I   M O U N T A I N S

SEWARD

Ninilchik River

NINILCHIK

KENAI
PENINSULA

Deep Creek

NIKOLAEVSK

ANCHOR POINT

Anchor R.

Resurrection Bay

HOMER

K E N A I

Kachemak Bay

SELDOVIA

Gulf of Alaska

NANWALEK

0   Miles   25

INTRODUCTION

"You can't be unobtrusive on three acres"

T HERE WAS NO MISSING THE "DRILL, BABY, DRILL" sign. It dominated the front yard of a house overlooking the north fork of the Anchor River and across the street from Armstrong Cook Inlet's natural gas operations on North Fork Road, about nine miles east of Anchor Point and 25 miles northeast of Homer by road. Surrounded by a scattering of children's outdoor toys and parked vehicles, it clearly stated this family's support of what was occurring near their home.

On that early spring afternoon in 2011 as I was passing through this sparsely populated area of the southern Kenai Peninsula, it was a reminder of the pro-oil and gas "drill, baby" gospel preached by Alaska's former Gov. Sarah Palin when she was the Republican vice presidential nominee during the GOP's unsuccessful 2008 campaign. She wasn't the first to proclaim it. Michael Steele, former lieutenant governor of Maryland who went on to become the GOP chairman in 2009, used the phrase during his speech at the September 2008 Republican National Convention in St. Paul, Minnesota. The crowd picked up the chant, their voices filling the Xcel Energy Center. Before that, "Drill, Baby, Drill" was the title of a blog posted by author and commentator Erik Rush. But it was Palin who put the phrase in the permanent vernacular when she hauled it out during a vice-presidential debate.

"The chant is 'drill, baby, drill.' And that's what we hear all across this country in our rallies because people are so hungry for those domestic sources of energy to be tapped into. They know that even in my own energy-producing state, we have billions of barrels of oil and hundreds of trillions of cubic feet of clean, green natural gas," Palin said in response to moderator Gwen Ifill's question to Palin and Democrat VP candidate Sen. Joseph Biden, asking if they supported capping carbon emissions.[1]

After her election as Alaska's governor in 2006, Palin's popularity had skyrocketed. In July 2008, shortly before her early resignation, she visited Homer to sign into law legislation sponsored by Rep. Paul Seaton, a Republican from Homer who as it happens is now related to me by marriage; he is my husband Sandy's son-in-law. House Bill 25 raised the standard of liability for landowners who allow public recreational use on their land. Paul said the legislation essentially came from avid skiers and snowmachiners in the Homer area and was so popular among Alaska's outdoor recreationists that it drew more testimony in the Legislature that year than oil and gas issues.

Working as a reporter for the *Homer News* at the time of Gov. Palin's visit, I was partnered with co-worker Ben Stuart to cover the public bill-signing event at the Alaska Islands and Ocean Visitor Center, headquarters of the Alaska Maritime National Wildlife Refuge. So eager were locals to see their governor that vehicles filled the parking lot and lined both sides of the Sterling Highway. Finding room inside the 120-seat auditorium was as difficult as finding a parking spot. Chairs were filled, spaces to stand were occupied, and the center's lobby was crammed with the overflow crowd.

In her comments, the governor praised the community for its work on the legislation. She also addressed the recent announcement that the U.S. Supreme Court had sliced an estimated 80 percent from a $5 billion settlement for more than 30,000 people harmed by the 1989 *Exxon Valdez* oil spill.

"I was hoping for a heckuva lot better decision than what we got. More of a remedy for coastal regions and fishermen. I have to apologize that we aren't in that kind of celebratory mood," she told the crowd. "Despite what happened to us, let us celebrate Alaska and Alaskans together."[2]

The following month, Palin was ready to leave the governorship behind in favor of a much larger arena. On August 29, Alaskans and the rest of the nation woke up to news that our governor had become Arizona Sen. John McCain's running mate. Reactions around the state were mixed.

"Friday morning's announcement stunned everyone, but here in Alaska, and particularly in the (Matanuska-Susitna) Valley, the stunned silence was quickly followed by, as Lt. Gov. Sean Parnell put it, a whoop of excitement," the Aug. 30 editorial in Palin's hometown newspaper, the *Valley Frontiersman*, stated. "And whether you are backing the newly revitalized GOP ticket of McCain-Palin, or not to be swayed by having Alaska's female top executive 'a heartbeat from the presidency,' the Palin pick is giving Alaska a positive spotlight like nothing else could."[3]

Larry Persily, who worked for three Alaska governors, including Sarah Palin, in the state's Washington, D.C., office and went on to become President Barack Obama's nominee as federal coordinator of the Alaska Natural Gas Transportation Projects, was a guest on NBC's "Hardball with Chris Matthews" in October 2008. Asked by Matthews if Palin should be vice president, Persily said, "No, I don't think she should be vice president. She's not qualified." He went on to categorize her as someone "immature politically" who "doesn't know what she thinks she knows."[4]

It was Palin's own words, including her rambling interviews with "CBS Evening News" anchor Katie Couric, that eventually raised eyebrows and lowered public confidence, mine included. When I saw the "Drill, Baby, Drill" sign in 2011, I could only shake my head at the lasting legacy of one of our state's most well-known personages.

. .

LATER THAT SPRING of 2011, efforts to bring natural gas to the Homer area were finally taking shape. Word of gas in the area was nothing new. In 1965, Standard Oil Co. of California, Richfield Oil Corp., Sunray DX Oil Co. and Union Oil Co. announced a dry gas discovery on North Fork Road, and in 1988, Mountain Alaska Energy announced it was making a run at bringing gas to Homer.[5] In 2010, after Paul Seaton put his weight behind bringing natural gas to the southern Kenai Peninsula, the Homer City Council added it to their list of legislative priorities. Then-Gov. Sean Parnell vetoed all but $525,000 of a hoped-for $4.8 million grant that year, but the city accepted that half million and contracted with ENSTAR Natural Gas to begin constructing what turned into a multi-year, multi-phase project. In 2011, when Parnell once again used his veto pen and redlined Homer's $10 million grant request, the city didn't lose heart and continued planning efforts with EN-STAR. The following year, Parnell finally approved an $8.1 million grant to build a pipeline from Anchor Point to Homer and Homer's tiny neighbor to the east, Kachemak City.

As part of continued coverage by the *Homer News* of the developing project, it made sense for us to interview people living around the well pad. Homer residents were being told natural gas would mean a cheaper, cleaner source of energy. Recalling the yard sign I'd seen across the road from the pad, I wondered if the project was equally good news for those living closest to the activity. With pen, paper and camera in hand, I drove back out North Fork

Road and knocked on doors, beginning with the home of Mark and Juley McConnell. Although their "Drill, Baby, Drill" sign was no longer in the yard, it seemed the most likely place to start.

The McConnells had purchased their half-acre-plus property in 2005 and moved onto the triangle-shaped piece of land in 2007. Mark said he and his neighbors were advised of the increased activity in their neighborhood through a letter from Armstrong announcing a public meeting.

"Of course there's noise when the (drilling) rig was going, but it's like any other noise, you get used to it," Mark told me during our interview.

"It's been loud, but not disturbing. Trucks go by, but the traffic is not any worse than it was, and for us it was OK because they kept us sanded out here," Juley said of much-needed attention paid to the road's icy winter conditions. "They've been really good to us."

It also helped that the project meant work for Mark, who was hired as a heavy-equipment operator for Starichkof, a locally owned service company working on the pad across the road from the McConnells' home.

Next door to the well pad was the home of Rick and Lori Paulsrud. A scraggly line of spruce trees did little to block their view of the drill rig and associated facilities and equipment. Neither was a distance of about 300 feet between the pad and their home enough to deaden the noise or soften the lights.

"We've lived here almost 11 years and if we'd had any idea when we bought this house that this was going in next door, we'd never have considered it," said Lori.

The side of the Paulsrud house facing the pad was windowless, which helped shield the couple and their son from the constant lighting.

"Probably the worst was when they were doing flare testing," Rick said, recalling "one time that (the flare) went on for about a month. . . . It sounded like you were listening to an airport with jets taking off."

With their dream of a close-to-nature, quiet Alaskan lifestyle destroyed, the Paulsruds found themselves considering a move. They found a home they were interested in buying and approached a Homer bank about financing contingent on the sale of their North Fork home and the three acres it sat on.

"They didn't want to do that with the purchase of this particular home, but (the bank representative) said she'd be willing to write up a proposal to the gas company to see if they'd buy our place," said Lori.

At the time I interviewed them for the *Homer News* story, the Paulsruds had not yet pursued that course. They had spoken with a realtor, however, and were told they would have a hard time getting out what they had invested in their home because of its proximity to the pad. If they could find someone interested in buying it, that is.

Those two vastly different perspectives haunted me. How could a community ignore the disruption its activity was causing a neighbor? Did the benefit to Homer residents justify the demolition of a family's dream?

· ·

FROM ANOTHER perspective and considering dueling reports, how had this project to bring natural gas to Homer come about? ENSTAR Natural Gas had said in 2008 that gas from Armstrong's newly drilled well, North Fork 34-26, would have an immediate use if taken north toward ENSTAR's already established markets in Anchorage—but if it was sent south, "you have to develop a system to accept it in Homer, as well as build a transmission line," according to Curtis Thayer, who was ENSTAR's director of corporate and external affairs at the time.[6]

"Homer . . . (has) a mature system, where gas lines would have to be built (under) sidewalks and driveways. It would make it a little hard to develop, but not impossible; just more difficult . . . and more expensive," Thayer said.

Ed Kerr, Armstrong's vice president of land and business development, told the Resources Committee of the Alaska House of Representatives in March 2009 that Armstrong was "very comfortable" that the North Fork prospect "held between 7.5 billion and 12.5 billion cubic feet of gas reserves" and said it was "realistic" that the prospect could hold between 20 billion and 60 billion cubic feet.[7] However, ENSTAR had said just a month earlier that there wasn't enough gas to add Homer to its distribution system.

"To serve a community like Homer, you have to have a minimum of 25 to 30 (billion cubic feet) to do it responsibly. You can't go ahead with all the required infrastructure changes and not have that supply in place. That would be irresponsible on our part," John Sims, ENSTAR's spokesman, said in 2009.[8]

Concerns also were voiced in 2009 that "sharp drops in gas deliverability from Cook Inlet gas wells" were causing the utilities and municipalities of Southcentral Alaska to brace for "emergency winter curtailments" of gas and power in Anchorage and nearby communities.[9] A contingency plan requested by the Regulatory Commission of Alaska and presented to the RCA by Brad Evans, CEO of Chugach Electric, envisioned a decrease of one third of the regional gas supply, with ENSTAR experiencing a 10 percent reduction.

A 2010 study by Petrotechnical Resources of Alaska indicated that gas reserves and production from Cook Inlet-area producers were declining at such a rate that "limited remaining development reserves in Cook Inlet and the long lead time required to bring new discoveries on-line, combined with the paucity of true gas exploration in recent years," meant that a source outside the area, "such as importation of other in-state reserves," would be required sometime between 2013 and 2016.[10]

"To meet demand by 2020, a total of 185 wells would have to be drilled," the report stated.

If I was reading it correctly, industry representatives were saying more natural gas was needed to meet already-existing needs, it would take years to meet those needs, and in the meantime Southcentral Alaska was approaching a shortage that would require bringing natural gas in from other sources. Further, ENSTAR said there wasn't enough North Fork gas to add Homer to its user base, but by Armstrong's numbers there was plenty. None of it seemed to make sense. The closer it came to a decision to bring the gas to Homer, the more the questions tugged at my attention.

Meanwhile, whether Homer benefited or not, more oil and gas activity was taking place on the southern Kenai Peninsula. On shore, gravel well pads and brightly lit drilling rigs began multiplying along the 52-mile stretch between Clam Gulch and Homer. The jack-up rig Endeavour-Spirit of Independence arrived in Homer aboard the heavy-lift vessel *Kang Sheng Kou*, destined to drill a couple miles offshore at Stariski, just north of Anchor Point. Buccaneer Energy was holding public meetings to discuss that project, as well as their West Eagle project 20 miles east of Homer.

An envelope addressed to my daughters and me that arrived in June 2013 brought the simmering questions and what I was observing to a boil.

"Hilcorp is interested in leasing your property for the purpose of oil and gas exploration," the letter from Hilcorp's landman David W. Duffy said.

Enclosed was a lease that, if we were in agreement, we were directed to sign. For the sum of $10 "and other valuable consideration, receipt of which is hereby acknowledged, and of the covenants and agreements herein constrained," our signatures would "grant, demise, lease and let unto" Hilcorp "the exclusive rights of conducting geological, geophysical, and other exploratory work for oil, gas, and associated substances, and for exploring, investigating, prospecting, marketing, drilling and operating wells for producing and marketing oil, gas and associated substances and producing, owning, saving, taking care of, venting, treating and transporting oil, gas, and associated substances therein, together with the non-exclusive rights to lay pipe, telephone and power lines, build and maintain roads and equipment, construct power stations, tanks and other facilities and structures, the right to remove all of said oil, gas, and associated substances, or water, or either of them, and the free and uninterrupted right and right-of-way into, through and under the said land at such points and in such manner as may be convenient or necessary for the purpose of conducting all operations for said oil, gas, and associated substances."

All of that. On 3.22 acres of land. For a term of three years "or as long thereafter as oil or gas and associated substances is produced in paying quantities from the leased premises hereunder, or lands pooled or unitized herewith, or drilling or reworking operations for oil, gas, and associated substances, are conducted thereon as hereinafter provided."

· ·

IN SIMPLER LANGUAGE, signing the lease would be signing over the 3.22 acres that are the claim my daughters and I share to my grandfather's, their great-grandfather's, homestead. The claim we have to an Alaskan heritage dating back generations. If Hilcorp struck it rich, how we would benefit was spelled out in a very complicated "royalty on production" section of the 12-page lease.

An updated version of the lease, including a Memorandum of Lease for recording purposes, arrived two months later.

I called Mr. Duffy and asked how Hilcorp proposed to do the work spelled out in the lease with my husband and me dividing our time between my husband's house in Homer and my cabin in Ninilchik.

"We'll be as unobtrusive as possible," was Mr. Duffy's response.

"You can't be unobtrusive on three acres," I told him.

I had a Homer attorney read over the lease. She advised against signing it, for two reasons. First, the royalty formula was so convoluted as to make it undecipherable. Secondly, she had seen bitter family disputes and divisions that tended to surface after a lease signer's death.

My daughters and I talked about whether or not to sign the lease. It was a short discussion. If either of them entertained visions of our becoming an updated version of the Beverly Hillbillies, the CBS sitcom family that left their backwoods home for Beverly Hills after oil was struck on their land, neither said a word.

As it had done and continues to do for many on Alaska's Kenai Peninsula, the oil and gas industry had come knocking at our door, asking our permission not to cohabit but to take over. For others, it has not knocked but simply established itself on a nearby piece of land. For some, as for the McConnell family, oil and gas has been a welcomed neighbor. It has robbed others, like the Paulsruds, of the value of their homes, obstructed views out their windows, flooded with light the dark nights once perfect for the Alaskan custom of gazing at the Aurora Borealis dancing overhead, shattered the peace that comes with building a home in less populated areas. And it has spread and strengthened the world's addiction to fossil fuels.

Drill, baby, drill? There is nothing unobtrusive about that philosophy, it seemed to me, no matter who utters the words.

My Roots I · A Foundation

— ❧ —

M Y PATERNAL GREAT-GRANDMOTHER WAS AN Alutiiq woman who traveled at the age of nine tucked inside the bow of a *baidarka* from what is now the village of Nanwalek at the tip of the Kenai Peninsula, up Cook Inlet to Kenai and eventually south to Ninilchik. My paternal great-grandfather, also born when Alaska was still a Russian colony, was of Russian-Native heritage. I, McKibben Autumn Diane Jackinsky, am the daughter of Walter Jackinsky Jr. and Alice McKibben, the granddaughter of Walter Jackinsky Sr. and Mary Oskolkoff, the great-granddaughter of Grigorii Oskolkoff and Matrona Balashoff. I have been shaped by my Alaskan lineage.

The history of my mother's people has had a less powerful influence on my life. But—born in Arizona and raised in Oregon—she left behind the familiar to join my father in Alaska. It is here that they raised their family and it is here, on a homestead originally claimed by my grandfather in the 1920s, that the rhythm of my heart was first aligned to the waves rolling along the shore of Cook Inlet, and here that my roots sink deep.

OUTSIDE MY cabin window this October afternoon, the view is of birch trees, their limbs stripped of golden leaves now that another fall is passing and winter looms. These trees have borne witness to my family's presence on the planet. Along with deep-green spruce and rough-barked cottonwood, they have stood sentry over the births of new generations and the passing of elders, over our prosperity and our poverty, our joys and heartbreaks. They have absorbed carefree laughter and voices raised in anger, lent their strength to children's swings and, when the cycle of life has brought them to Earth, filled our stoves, warmed our cabins, produced blazing campfires to light the darkness.

Today's scene reminds me to look, not with the limited vision of my eyes but with the future-seeing, hopeful gaze of my heart. Once-green leaves carpet the ground to become another layer in the soil. Cottony fluffs of fireweed seeds hug tall, skeletal stalks and then are carried off in the wind. The musky scent of highbush cranberries floods the air with the message that now is the time to harvest. Be steadfast, the chilly morning air tells me: The fullness of life will come again. The transformation from summer's bounty to winter's bareness is simply a step in a cycle. Rebirth will come once winter has passed and the northern hemisphere tips again toward the warmth of the sun.

However, another sort of energy has found its way to my front door: the growing momentum to discover additional oil and natural gas fields. Cleaner energy than the wood from these forests and coal from the beaches. Cheaper energy than the heating oil delivered by fuel trucks. Testing might discover it is here, beneath my feet. It could, as it has for neighbors to the north, provide a source of income beyond what I thought this land could provide. It could change everything.

All I need to do is what I'm told my neighbors have done: Sign this piece of paper. And then step aside, as the land that has been in my family for generations becomes unalterably changed into an unavailable, unfit, nonexistent haven for future generations.

· ·

Ninilchik began as a settlement in the 1840s for pensioners of the Russian-American Company and their Native or Creole wives and families. After arriving in Ninilchik about 1912, traveling from Vilnius, Lithuania, by way of the East Coast, my grandfather Walter met and married Mary, a young Russian-Alutiiq woman born in the village. The home where my grandparents' six children were born was constructed in the village style, hand-hewn logs, each expertly cut to fit the one beneath it by John Ostrogin, the builder of most village homes at that time.

When my father was about 10, Grandpa picked out a piece of property three miles north of the village to homestead. There, John, who also was known by the Russian name Vanek or Peg-leg John for the wooden leg he fashioned to help overcome a disability with his left leg, built the family a second home. With a root cellar under its floor, the main room was approximately 16 by 16 feet. Off the back, a kitchen was added. Off the right, a single room served as a small bedroom. Above the ceiling, an attic provided sleeping space for the children.

A warehouse was later added, but it came toward the end of Vanek's life, when his health was failing and his home-building artistry had passed its prime. The 20-by-20-foot one-room structure provided space for Grandpa to store what was needed for the fox farm he developed, tools to maintain a garden on the south-facing hillside behind the house, and a place for coal hauled from the beach to keep the house warm.

My father, who died in 2012 a few weeks short of his 96th birthday, recalled how he, his brother Edward and their father took turns making the six-mile round trip from the homestead house so the foxes could be fed. No matter the weather, and even when the family was in residence at their house in the village, the daily journey had to be made, by snowshoe, sometimes by dog-sled, sometimes on foot.

In his 20s, the homestead house was home to my dad, his first wife, Marie von Scheele, their daughter, Vonnie, and their son, named Adolph after Dad's younger brother, who had died in a plane crash on his 16th birthday. They shared it with my grand-father until Dad and Marie's marriage ended a few years later.

Dad and my mother were married in 1947, and after my birth in Oregon in February 1949, the homestead house was our home for a time. When Grandpa died in 1950, the homestead was di-vided among Dad, his brother Ed, and their sisters Margaret and Cora. Dad's portion included the house. When I was a toddler, my parents built Jackinsky's Ranch, a combination bar, liquor store and living quarters on another portion of Dad's land, and we left the homestead house.

Recollections of living there are long gone, but I remember vis-iting the empty structures—the house and the warehouse—often. The dampness of the unheated, vacant rooms. The thick smell of moose meat hanging from beams in the warehouse after one of Dad's successful hunts. In the summer, the garden drew us back to that plot of land. The outstretched arms of strawberry plants created a netting that since my father's childhood had spread across the hillside. The log and chicken-wire fence built during Dad's youth did little to contain their advance. The green of the plants was a welcome sight after winter. Then came the white blooms and finally the plump, sweet berries that we hurried to pick before greedy little rodents nibbled them down to the stems.

The garden also contained gooseberry bushes that Dad's brother Adolph had planted. The tight, round, luminous green berries lacked the lure of sweet strawberries, but their connection to a family member I never met made them memorable. When my family moved to a home on the south side of Ninilchik, the bushes, as well as some strawberry plants, were dug up and taken with us.

Just below the bluff from the homestead was the stretch of Cook Inlet coastline where our family fished commercially during the summers. We moved into the small plywood one-room cabin—painted orange on the outside—shortly after school let out in May and were there until shortly before school began again in the fall. Early in the spring, a pile-driver would arrive from the cannery to help set poles for a fish trap. Its rhythmic pounding marked the beginning of the season. In addition to tending the trap, Dad drift-fished with a gillnet on Cook Inlet, while Mom, my brother Shawn, sister Risa and I tended set-nets close to shore.

When we weren't fishing, the beach offered plenty of opportunities for a child's imagination to run wild. We had the bluff to climb, driftwood to pile into forts, beachcombing to collect interesting shells and rocks. There was frequently a fire on the beach in the evening, the sweet smell of cottonwood bark that had washed up on the shore adding its perfume to the salty air. A chunk of fresh salmon with slices of onion wrapped in foil and placed in the coals of the fire made a delicious supper.

Inside the cabin, two sets of bunk-beds provided sleeping space. I had a top bunk and Mom occupied the bunk below mine. At the foot of her bed, a crosswise "shelf" provided a sleeping area for my little sister. In the other set of bunk-beds, my brother had the top bunk and Dad anchored the bottom.

Sand was a constant companion, tracked into the cabin on our shoes and sifting from our feet into our sleeping bags. Meals were cooked on a kerosene camp stove. A wooden box nailed to the outside of the cabin served as a "refrigerator," the food chilled by cool air off the inlet.

Salmon was the mainstay of our Ninilchik diet in all seasons. Smoked, canned, pickled, frozen. In soups, in loafs or patties, in sandwiches, as steaks or roasted. We dug razor clams on Cook Inlet's minus tides. Moose meat was our "beef."

Vegetables that the garden didn't provide, as well as fruit, butter and milk, came in cans from the cannery at the end of each fishing season. The day those cases arrived was like Christmas, as we unpacked the boxes and filled our shelves for the winter months ahead.

· ·

PUTTING TO USE his World War II ocean-going experience in the U.S. Merchant Marine, not to mention a lifetime on the water as a commercial fisherman, my father went to work for the Alaska Marine Highway System when its first ferry, the M/V *Malaspina*, was launched in 1963. The following summer, my family moved away from Ninilchik to Juneau, the ferry system's homeport at the time.

There were return trips to Ninilchik to visit family and friends during the summers, and one year Mom made the journey with us kids for the Christmas holidays. But the village setting, its mostly subsistence way of life, and the family we'd been surrounded by were at a great distance from our new city life.

Distant it may have been, but never far from my heart. Increasingly, as a young wife and then the mother of two young daughters, I felt the pull to return to Ninilchik and the simpler lifestyle of my memories. But my husband's family was in Juneau, as was his work and the life he and I were creating for our own family.

My mother had since returned to Ninilchik, however, and had set about building a cabin for herself on a piece of the homestead that offered a wide-open view of Cook Inlet. She carved a road out of the forest to her chosen site. As she worked constructing her small frame home, she slept in an aluminum storage shed, carried water, cooked on a small two-burner stove and fashioned an outhouse in a nearby patch of alders. Accepting little help from others, she eventually and proudly moved into her 14-by-16-foot cabin.

In 1978, when my own marriage began unraveling, it was to Mom's cabin I took my two daughters, Jennifer and Emily, and myself. Not only was I beginning a new chapter as a single parent, but I was beginning from scratch. Mom made room in her tiny living quarters for bunk-beds we built for the girls, who were 6 and 5 at the time. We had a Franklin stove to provide heat and kerosene lanterns for light. Meals were prepared on the two-burner kerosene cookstove that had served her in her aluminum-shed days.

The quarter-mile dirt driveway was frequently impassable because of mud, and water still had to be carried in, in five-gallon jugs. We took showers and did our laundry at a Laundromat in Anchor Point, 20 miles to the south.

· ·

EMPLOYMENT opportunities are scarce in Ninilchik. The summer offered cannery work, however. So, from early in the morning until late at night, I stood in ankle-deep icy cold water, stripping the back bloodline from commercially caught salmon that slid past me on the processing table. It paid $3.50 an hour, and without my mother's help with the girls I'd never have managed.

Mom's generous hospitality and unfailing support were appreciated, but I recognized that a long-term living arrangement had to be found. That's when my eyes fell on the still-standing main room of the old homestead house.

In the years since we'd left Ninilchik, Dad had sectioned off pieces of his portion of the homestead for my mother, my brother, my sister and me. The old house sat on my brother's piece. With the structure lacking any kind of foundation other than the earth beneath the bottom round of logs, moisture had done its work and rot had taken hold. Shawn didn't want the house and planned to tear it down. If I promised to move it, could I have it? His answer was yes.

My first step was to sketch it on paper, log for log, numbering each log on the paper and then correspondingly on the logs themselves. Once that was done, friends Patsy and Steve helped me begin the process of removing the roof, which, we were startled to discover, was home to several bats. Dismantling the logs one at a time came next.

It was thrilling to determine that all but the bottom two rounds were sound. It also was a process of stumbling across unknown bits of family history. A skinning board used for fox skins helped reinforce the ceiling. A piece of sock, a strap from a pair of coveralls and other pieces of clothing were stuffed between the logs, along with layers of moss, for insulation. Pages from old newspapers, also adding insulation, had to be peeled from the walls. We found Uncle Adolph's name carved into one of the logs.

Once the structure had been taken apart, we carried all the logs to my piece of the homestead and stacked them for reassembling once I finished a piling foundation and construction of a floor.

Patsy and Mom helped with that endeavor. With a how-to book as a guide, we laid out the perimeter, marked the position for the eight-foot-long piling, and began digging. I purchased the poles from Homer Electric Association, to be delivered to the end of the driveway. Another friend, Suzi, helped us move them in the back of a pickup to the building site.

If you've ever tried to lift a wooden utility pole, even just an eight-foot section, you know how heavy they are. None of us had had that experience, and we were shocked to discover new limits to our strength. But we were not defeated. Setting up a series of blocks, we slowly raised the end of each pole, one at a time, until it finally reached the level of the pickup's bed. With that supporting one end, we had the leverage we needed to raise the other end and shove the pole into place.

Through the summer and fall, we continued our work on the floor, constructed two new rounds to replace those that were rotten, and slowly raised the walls. By widening the front door and the entry that once led to the kitchen, I enlarged the cabin to a 20-by-16-foot dimension. The entry into the no-longer-existing bedroom was framed in to become a window. The final placement of the top wall log provided a satisfying "whoosh" as it pulled the house into shape, perhaps testifying not to carpentry skill but to luck.

Throughout the summer, my efforts frequently met with criticism from Ninilchik's male population. "Women's liberation" may have been erasing limits in other locations, but the movement had yet to make much of a difference in Ninilchik. One afternoon, Harry, a local who had grown up in the Ninilchik of my dad's generation and who had been one of the biggest critics of my efforts as a single woman with two daughters, stopped by to see how the building project was coming.

"I wouldn't have believed this if I hadn't seen it!" he exclaimed, with no idea how satisfying his words were to hear.

· ·

As FISHING began to wind down at the end of summer, I found work at the Inlet View, a local restaurant where I took orders and cleaned tables, under the supervision of cooks Maggie and Edna. Ninilchik's lack of a kindergarten for Emily prompted me to organize a preschool, with funding from Ninilchik Natives Association. It met two mornings a week, and I had a third morning to do prep work. I also wrote a column of Ninilchik items for a small weekly newspaper in Homer, the *Homer News*. That, plus the financial help I received from the girls' father, made it possible for us to survive that first winter.

WHILE IMPRESSIVE progress had been made on rebuilding the homestead cabin into a place for my daughters and me, it still lacked a roof. Work continued until the weather brought construction to a halt. As hospitable as Mom was, the strain of the four of us living in that small space finally pushed me to find other housing. First, Jennifer, Emily and I rented a two-story cabin owned by some friends. Then, in February 1979, on a weekend when the temperature had dropped to minus 26 and an icy wind was blowing down the Ninilchik River valley, we moved into a small two-story cabin in the village. My reasoning—that the girls would have the experience of living in the village—overrode my good sense.

Our first Monday morning I awoke to black soot from the oil cookstove lining the creases of my face. The inside temperature hovered in the 50s. Dressed in our snowsuits, we ate lukewarm, partially cooked oatmeal. The car was frozen. Jennifer and I were crying. Emily, the youngest in our little trio, kept reassuring me: "It's OK, Mommy."

Somehow I managed to get Jennifer onto the school bus. Then I ran to the neighbors to borrow a phone and call one of the preschool parents to ask for a ride. As happy as I was to see Greg when he arrived to pick Emily and me up, his first words made me cry even harder.

"Jesus Christ, you can't live like this," he said.

After that morning's preschool session, Emily and I bundled into our snowsuits again and began the two-mile walk back to the cold house. She was a brave little trouper, but, headed into the face-numbing weather, repeatedly asked to stop. Fortunately, our friend Steve came along and not only drove us home but also showed me how the stove's carburetor was clogged and in desperate need of cleaning. That made all the difference. Suddenly, our home began filling with warmth.

We continued to live in the village through the winter. The stove was a constant problem. The sink had running cold water on a good day. The toilet worked infrequently. But we were on our own. And we could enjoy the hospitality of neighbors who had a *banya*, a steam bath, that we were invited to use when their sessions were over.

There also was the Russian Orthodox Church on the hill overlooking the village that my father had attended when he was a child. Although we didn't attend services, the sound of the bell on Sunday mornings was a reminder of my family's place in the village and a touchstone to my own determination.

"Adversity builds character," my mother frequently recited to me. I knew the truth of what she was saying, but I often found myself tired and impatient with the adversities day-to-day living offered.

. .

ON MEMORIAL DAY weekend, with the close of school for the year, my not-yet ex-husband arrived to take our daughters to his home in Anchorage for the summer. Rather than return to the cannery, I crewed on a halibut boat on Cook Inlet and then worked as a cook on a drifter.

Throughout the summer, I continued to work on the cabin. By the time the girls returned to Ninilchik for school, it was finished enough for us to move in. My fishing money helped purchase a wood-burning Earth Stove for our source of heat. Some simple wiring offered a couple of electric lights. A propane stove gave me the luxury of four burners and an oven for cooking.

We lacked running water, but the outhouse was positioned so we could see if company was coming, and in the evenings the Big Dipper was bright in the sky to the north. Laundry still had to be done at a Laundromat, but we had baths at home, one at a time, with water heated on the stove and poured into a galvanized washtub. A three-quarter loft made a warm sleeping area for the girls, and a small addition on the back of the cabin gave me some privacy.

Visqueen stapled over the single-pane, recycled windows on the main floor provided added insulation. A double layer of Visqueen in the loft covered openings my limited finances failed to fill with glass. (During one particularly strong fall storm, I was awakened by Jennifer and Emily screaming for me from the loft. Racing up the stairs I saw that the wind had ripped the Visqueened windows open and was quickly stealing our heat, while Jennifer and Emily clutched their sleeping bags to their chins.)

Styrofoam cut to fit the seat of the outhouse provided a warm surface against bare skin, but not enough to make the girls comfortable. Their choice, especially during the night, was a bucket to be emptied in the morning.

After their return to Ninilchik, I decided to homeschool them rather than re-enroll them in public school. We used a program of study provided by the state of Alaska. Participating in their education from that front-row seat was one of the highlights in the time the cabin was our home. A time that was all too short.

With fishing season over, I'd begun working as a checker at a grocery store in Kenai, 50 miles to the north. Mom watched the girls on days I worked. Their studies were fit in around my schedule. It seemed like a workable solution, and I was thankful for the employment opportunity. However, on a day off, on our way home after time spent in Homer with friends, I fell asleep at the wheel of my Datsun B-210, and that brought the commuting to an end. Fortunately that's all it ended, since none of us were wearing seatbelts, but the car was totaled, and, lacking transportation, I had to give up my job. Desperate to find work, and with limited local opportunities, I called a former boss in Juneau and was told there was an opening. If I could get us there, I could have the job.

And so we began yet another chapter of our lives. The smell of wood smoke from the cabin was finally washed out of our clothes, and I gradually became used to turning up a thermostat for more heat rather than having to add a log to the fire. The ease of indoor plumbing was a welcome change. But the desire to return to Ninilchik, to the homestead and to the source of all that was familiar in my life, was not so easily erased.

EMIL

"I couldn't take it with me"

TWELVE MILES NORTH OF NINILCHIK, THE Sterling Highway dips down past a low-lying swampy area dotted with
scraggly bog spruce. A small creek runs through a culvert
under the highway, carrying water away from this spongy wetland
to Cook Inlet a few miles to the west, where the scene opens onto
a broad view of the snow-covered peaks of the Aleutian Range
rimming the blue waters of the inlet's distant shore.

To the east, the taller relatives of these scrub spruce and the
birch that share the forest beyond have been cleared back to ex-
pose the right of way under which runs another sort of culvert, a
high-pressure pipeline that carries natural gas past the scattered
homes and businesses along the highway to more densely populated areas of Southcentral Alaska.

In my mind's eye, as the highway climbs again a gentle hillside
slopes away toward the west, its top crowned with a stand of birch
trees. This marks the 160 acres on which the Bartolowits family
built their home in the late 1940s.

I remember coming here as a child with my parents when the
Bartolowits house was under construction near the top of the hill.

The family lived in below-ground-level quarters that would become the basement of the finished house. While my mother and father visited with Fran and Emil, their daughter, Tessa, and I played with her collection of dolls.

My own family of dolls was loved, but limited. Tony had wild platinum blonde hair that my attempts to brush and rearrange only resulted in an unruly, tangled mess. The clothes she originally wore had long since disappeared, replaced by a wardrobe I'd fashioned from discarded socks. Another doll, whose name I don't remember, had a soft, cloth, easy-to-hug body and a head, arms and legs made of something sturdier. Not so sturdy, however, that they could withstand the impact of falling down the ladder from the attic bedroom my brother and I shared at the time, above the living quarters for "Jackinsky's Ranch," the bar and liquor store Mom and Dad operated three miles north of Ninilchik village. As a result of the fall, one of the doll's legs had cracked and her eyes, designed to close as if in sleep when I laid her down, had disappeared inside her head. Still, she was treasured.

But Tessa's doll collection was something altogether different. Tessa's dolls were beautiful, most of them to be admired from afar, as untouched as the day they'd been purchased.

In other ways, too, visiting the Bartolowitses was like opening a door on a different world. Root cellars under the floors of Ninilchik houses were the perfect place to store a summer's worth of vegetables. A basement that provided additional living space was beyond my imagination, and Tessa's dolls stored in boxes and stacked neatly on shelves hinted at unbelievable abundance. My eagerness to look inside those boxes mirrored my eagerness to reach out for all the luxuries the Bartolowits family seemed to offer. On the other hand, I also found security in the void that I sensed separated what the Bartolowits family, from a world outside of Alaska, represented and the world I knew, which gave definition to my family and our Ninilchik village way of life.

In the decades that have followed, passing the driveway leading to the long-since-finished Bartolowits home has brought a smile with those childhood memories.

Until one fall day in 2013. The view of the inlet and the mountains beyond was still there, but the hilltop to the west had been bulldozed and was crowned not with birch trees but with a large gravel pad. On it stood Atco units, yellow and white trailer-like facilities, along with pickups and pieces of heavy equipment, some parked with their motors left idling.

From the center of the pad rose the tall derrick of a natural gas well.

The change was more than jarring. Tears blurred my vision, and without realizing what I was doing I slowed the car as I took in the activity replacing the once gentle landscape. "What happened to the Bartolowits homestead?" I asked my neighbor John at the post office a few days later.

"Emil sold it," he said, shaking his head.

I began pouring out my disappointment and heartbreak over what had been done to a once-beautiful piece of land, but John cut me off.

"I was sad, too, but then I had to realize that I drive a truck that uses fuel," he said. "I guess I can't be too critical."

Several months later, sitting in the living room of his new home, Emil told me about his decision to sell.

· ·

BORN IN AUSTRIA, Emil came to the United States in 1929, when he was eight years old. His father had made the journey several years earlier, after the end of World War I, and Emil's mother had followed, leaving Emil and his brother with their grandmother until their parents could send for them. The two young boys were accompanied to the United States by a 21-year-old woman who took care of them on the trip.

"I remember going into New York Harbor and seeing the Statue of Liberty," Emil said.

The family settled in Pittsburgh, Pa., where he began school as a second-grader not knowing a word of English.

As a young man, Emil served in the U.S. Marines for six years. Shortly before being discharged, with his uniform in need of some repairs, he visited a tailor shop and heard how overworked the owner was.

"I said, 'Heck, I'm going to be discharged in a week or two and will be looking for work.' He said, 'If you want to work, I'll sell you the place,'" Emil recalled.

Although he knew nothing about the tailoring business, the idea appealed to him. A deal was reached that included the owner teaching him the business and working in the shop while Emil and his new wife, Fran, took their honeymoon.

It was while he was operating the tailor shop that the idea to come to Alaska began taking shape.

"I had the radio on all the time and kept hearing 'Go to Alaska, the last frontier, and you can get 160 acres for homesteading.' I heard that for about two weeks and got to thinking it might be a good idea," he said. "That night, I asked Fran, 'What do you say we go to Alaska?' She said, 'Well, when do we go?'"

The Homestead Act, signed into law by President Abraham Lincoln in 1862, made 160 acres in one of the western states or territories available to a person 21 years of age or older or the head of a household or someone who had served in the military. The person had to live on the land for five years, develop it for agriculture and build a house on it. If those requirements were met, at the end of five years the person was granted full ownership of the parcel. Although provisions of the Homestead Act were extended to Alaska in 1898, it was after World War II and the Vietnam War that homesteading in Alaska saw an increase.[1] Railroad construction in the early 1900s and completion of the Alaska

Highway during World War II also added to the attraction of homesteading in Alaska.[2]

In 1947 Emil and Fran headed north, driving a 1936 Ford panel truck and towing a 16-foot-long, eight-foot-wide trailer Emil had built.

"We sat down and made out a menu of what we needed per day and then consolidated it into cases of this and cases of that and that's what we had for food," said Emil of the year's worth of groceries they put together. "We were set."

The trailer was filled with furniture, and the back of the Ford was filled with groceries. The space between the cases of food and the top of the vehicle was just enough for Emil and Fran to sleep there while on the road.

In Anchorage, Emil went to work for a business that made pumice stone blocks using pumice from Mount Augustine, an island volcano in the lower Cook Inlet.[3]

Emil's co-workers included Per Osmar, Floyd Blossom, Bud Herman and others who were making plans to homestead in the Clam Gulch area of the Kenai Peninsula. They encouraged Emil and Fran to join them.

"The barge that hauled the pumice stone from Augustine owed us wages, so that was our fare coming down on that barge," said Emil.

The band of homesteaders was dropped off on the beach and set about identifying the boundaries of their property.

"I'm the only one that filed and stayed on the original filing," Emil said.

It was 1953 before the couple made a trip to the Lower 48.

"That's because Fran's mother came up and she paid our way," he said of the long drive and the cost involved. "Then it was another four or five years, I forget how many, before we went out again."

Opportunities to make a living on the Kenai Peninsula were limited, but Emil was determined to do whatever he could to be close to his family. He fished commercially for John Huey for a time and then, in 1954, bought his own fishing sites for $1,000.

"The first year I didn't make enough money to pay the $1,000, but it was OK," he said of the binding handshake that had sealed the purchase.

The first Bartolowits home was a log cabin. The family grew to include daughter Tessa and three sons, one of whom died, Emil said, when he was about three years old. With no doctors or hospitals in the area, when Fran was due to have their first child Emil borrowed a car and took her to Seward, some 120 miles across the peninsula, where the nearest hospital was located. Leaving Fran with friends, Emil returned to the homestead. In the absence of telephones, and long before email and cell phones, he learned through Mukluk Telegraph, a program on an Anchorage radio station that broadcast personal messages of all sorts for people in outlying areas, that he had become the father of a baby girl.

By the time their sons were born, there was a doctor in Soldotna, a comparatively short 25 miles away, but no hospital, so the Bartolowitses arranged for what Emil described as at-home nursing care.

Looking back on those years, Emil recalled their self-sufficiency and simplicity.

"They were the best times of our lives," he said. "We had nothing. We had no worries. We had no money, but we didn't need any. It was a good life."

When he and Fran moved to Alaska they told themselves they would stay at least two winters.

"And, of course, it turned into 60-some years," he said.

In the early 1960s natural gas was discovered just north of the Bartolowits homestead, and over the years interest in the area from an assortment of oil and natural gas exploration companies continued to grow.

The summer of 2013 brought huge changes in Emil's life, most significantly the loss of the person with whom he'd shared more than 66 years. Fran died on June 17.

Also during the summer, the couple's son, Paul, a production foreman for Hilcorp Alaska, received a call from someone with the company asking if he knew an Emil Bartolowits.

"They asked him if I'd be interested in selling the place. They wanted 10 acres. He said he didn't know, they'd have to ask me," said Emil. "So they came down a couple of weeks later and told me what they wanted."

Emil's immediate response was that he would have to discuss it with his children, but he also began weighing the pros and cons of a different lifestyle, one that didn't involve the stairs in the home he and Fran had built.

"I'm 92 years old and I have no problem walking up and down stairs, but I thought, well, hell, who knows? So we discussed it and I thought if the price was right, I'd sell," said Emil.

Some of the homestead had already been given to his kids. Some of it was a swamp that Emil knew he'd never be able to sell by itself. So he made a counter-offer.

"I told them, 'Well, if you want it, you'll have to buy it all,'" he said of a deal that included setting aside a two-acre parcel for himself.

The sale price provided enough for Emil to have a new, one-story house constructed. He also purchased a new vehicle.

With Hilcorp eager to have access to the land, Emil was fortunate to locate a builder who immediately began construction. On July 4th, less than a month after Fran's death, he had a garage sale and, with the help of family, began clearing out personal belongings as well as larger items from the house in which they had lived.

"We even sold the furnace. . . . You can't believe the stuff we sold. It all went. Sixty years worth," he said of the day's effort, which netted more than $12,000.

By the middle of August Emil was in his new home. Rather than facing west, toward Cook Inlet and the mountains, it faces east, toward the rising sun. And the swamp. It also looks out on the place where he and his family once lived.

"I could see my house from here and then two hours later, it was gone," he said of witnessing the end of that period of his life and the beginning of Hilcorp's exploration for natural gas.

All his dealings with Hilcorp have been positive, he said.

"They cut the hill down about 16 feet and always asked if they could do it, if it was OK. They had already bought it, but they were very considerate," said Emil. "I can't say anything bad about them."

In the quiet of his living room on the winter afternoon we visited, Emil talked about how his decision has benefited others.

"Believe it or not, because I sold, it put a lot of people to work. Local people," he said, listing construction companies, heavy equipment operators and surveyors. "I don't know why (Hilcorp was) bound and determined for that parcel. ... But they felt pretty sure of that. So why should I hold it up? Hell, the few years I'm going to be here, let them continue."

There was a time when it was commercial fishing that put money in the pockets of Kenai Peninsula residents and food in their cupboards, "but I think oil and gas is what makes this peninsula work," said Emil. "I may be wrong, but it employs a lot of people. They've got lots of gas if they can find a place to use it."

Where it won't be used is in Emil's house. Any natural gas produced on the Bartolowits Pad, as it has been named, will be added to what's flowing through the pipeline that runs less than a mile from where he now lives, destined for other locations.

The house in which I visited with Emil is a long way from the unfinished house I remember visiting as a child. This one, in a short amount of time, is completely finished. No exposed two-by-fours, no plywood floors. It is roomy, with large windows letting in the late afternoon winter light, and it is comfortably filled with a mixture of new furniture and reminders of another time, when the sounds of children filled Emil's life and Fran was never far away.

Was it hard to let the land and house go, I asked him.

"Yes, it was hard, but I knew I couldn't take it with me. That was my feeling," he said. "And my kids have their own places. They're established wherever they are. So we agreed that was the best."

What about changes to the environment, I asked, thinking of the hill he and his family once lived on. In response he told me about one tree that had a bald eagle nest in it, and praised Hilcorp for a commitment to maintaining a 300-foot distance from the tree.[4]

"This environment stuff, it has its place, but (environmentalists) go overboard, in my estimation," Emil said.

Finally I asked how he felt about the work being done and the money being spent to explore for a resource that will not benefit him in his new one-story house on his two-acre parcel. In his answer, I catch an awareness of less favorable outcomes and possible consequences.

"You figure what they drilled and the effort they put into that land, it's millions of dollars. That's a gamble," he said. "When it pays off, it pays off good. It's got to because when they have a spill, who is going to pay for it? They are. They have to have that money in the bank."

ROBERT AND STACY JO

"It felt like an alien invasion"

HOW DID THOSE LIVING NEAR EMIL BARTOLOWITS react to the sale of his property for Hilcorp Alaska's expanded search for natural gas? I didn't have far to look for an answer. The "For Sale" sign on the opposite side of the Sterling Highway was a good clue to what I'd find when I stopped on a chilly afternoon in November 2014 to introduce myself and ask if someone would be willing to talk with me.

A gray-haired man I guessed to be near my age pointed me toward a younger man a short distance away, who with the help of a second gray-haired man was piling items onto an open trailer. Judging by the loads already filling the backs of two pickups, a move was clearly in progress.

They were hard at work, carrying items from a barn and securing them to the trailer, but as soon as I explained I was writing a book about the impact of oil and gas exploration on southern Kenai Peninsula residents, Robert Correia turned his attention to me. Without any prompting on my part, he immediately and passionately began sharing his family's experience, while his two companions, his father and father-in-law, nodded their agreement.

The items filling the two pickups and the trailer, the stacks of belongings on the frozen ground awaiting packing, the barn, the corral, the house—all belonged to Robert and his wife, Stacy Jo. Since purchasing the seven and a half acres in 2007, they had built this home for their family, perfectly fitted to their lifestyle. It was, they had assumed, where they would live while they raised their family and on into their retirement years. What was happening with Emil's land and the resulting impact on their property had shattered that dream.

"Before, we could drive up and down our driveway and see the trees, look down Bartolowits' driveway and see the beautiful view," said Robert, his raised voice filled with anger as he looked across the highway at the gravel pad littered with pickups, heavy equipment, office trailers and a well head. The spot once featured birch trees that framed a view of Cook Inlet and the mountains on its distant shore. But those trees and much of the soil had been removed, making room for the equipment necessary for Hilcorp's drilling operations.

"Now you can see 'the view' from everywhere, but it looks like a war zone," said Robert.

A few days later, I sat with Robert and Stacy Jo in their family's temporary quarters, a bed and breakfast six miles from where they had dreamed of spending the rest of their lives, and listened as the couple described what had driven them from their home.

Robert and Stacy Jo are both the third generation in their families to make Alaska home. Robert's grandfather traveled north from California in 1950. He found work in Anchorage and eventually homesteaded on the Kenai Peninsula. In the early 1960s, Robert's father began spending his summers in Alaska. After graduating from high school in Michigan, he made the permanent move to Alaska. Robert's childhood was spent commercial fishing with his Alaska family.

"I'm a third-generation Cook Inlet drifter," he said, clearly proud of making his living from Alaska's cold, pristine waters. The inlet, 85 miles in width where the North Pacific pushes its way in, and becoming narrower as its funnel shape cuts more than 150 miles into Alaska's southcentral coastline, is infamous for its strong currents and tidal action. Turnagain Arm, the inlet's northeast corner, claims the largest tidal range in the nation, with a mean of 30 feet, and ranks fourth highest in the world.[1] Robert's introduction to fishing was as part of the inlet's fleet of commercial drift fishermen who sell their wild-caught salmon to local canneries for distribution around the world. In 2005, he expanded his fishing experience to include fishing set nets that extend from the inlet's shoreline.

Originally from Texas, Stacy Jo was two years old when her family moved to Kenai, Alaska, in 1974. She met Robert in 1994, they married in 1995, and their family now includes six children.

Robert and Stacy Jo established a home base for their family on a hillside just off the Sterling Highway. It offered a sweeping view of the Kenai Mountains and the vertebral horizon shaped by their peaks. As rich as it was in view, however, it lacked flat ground to store Robert's collection of nets, reels and boat engines as his fishing venture grew.

"It had only about a third of an acre you could live on and have stuff parked on. With all the fishing gear, we were kind of squished," he said.

As Robert drove along the highway one day his attention was captured by a "For Sale" sign. He stopped, got out of his vehicle, walked around the site and said to himself with absolute certainty, "This is where I want to live."

After purchasing the property that spring, the Correias immediately began building. A driveway was constructed by June, space for their home was cleared by July, and, as soon as the summer's busy fishing season was over, a cement slab was poured for the foundation of their 32-by-40-foot home. By the following February, enough of the structure was finished that they could move into the downstairs while continuing to complete the rest of it. In 2012, they added the barn and corral.

The heavily wooded piece of land gave the family plenty of privacy. The tall, thick-limbed spruce and birch trees shielded them from the sights and sounds of highway traffic. There was ample room for the horse barn and corral, the steers and cows, the chickens, turkeys, geese and rabbits, the family's dogs and cats. It put Robert closer to the site of his summertime fishing operations and, equally important, provided wintertime storage space for all the trappings of his chosen profession.

"We had a narrow driveway (from the highway) and then it opened up and it was our own little world," said Stacy Jo.

When they learned of a 3.5-acre piece of property for sale next door, they purchased it, too, "in case we retire," said Robert. Pooling their fishing income with Permanent Fund Dividends, they were able to pay for the $15,000 piece of property.[2] It was, Robert told me that night at the B&B, something they never would have done had they known how their lives were about to change.

· ·

IN JULY 2013, the Correia family stopped at a garage sale being held by their 92-year-old neighbor, Emil Bartolowits. It was more than a garage sale, they discovered. Emil was selling everything, including his home, piece by piece.

"Windows. Doors. Everything," said Stacy Jo. "We were surprised."

Not long after the garage sale, Hilcorp stripped away the trees, leveled the knoll, and began setting up drilling operations.

"(Emil) had a beautiful lawn and they just covered it all up with gravel, and I thought to myself what a waste of that beautiful lawn he'd cared for," said Robert. "And there, right there between the house and the highway, were some big beautiful birch trees. ... The trees came down and the (drill) rig went up."

The death of that stand of birches was a particular blow to Stacy Jo. In the six years the Correias had been Emil's neighbors, Stacy Jo had come to eagerly anticipate the welcome the outstretched branches offered each time she returned home. No matter the time of year—whether adorned with springtime buds, the full leaves of summer, the golden glow of fall or the sparkling snow of winter—their greeting was a sign she was home. The day Hilcorp cut down the birches is a scar burned into her memory.

"I actually put a chair at my window and watched them come down," said Stacy Jo, the sorrow of that afternoon evident in her features. "Out of respect for all the joy they brought me, when I came home and saw them I pulled up a chair and watched."

A line of spruce trees and alders bordering the highway side of the Correias' property helped soften the visual impact of the devastation that was occurring. Or at least it did until the Alaska Department of Transportation and Public Facilities scraped away the right-of-way vegetation. According to the department the goal was improved visibility and safety, particularly when it came to avoiding collisions with the area's moose population. With that foliage gone, however, there was nothing to shield the Correias from the changes being made across the highway, nothing to block the glare of the well pad's Halogen lights flooding Robert and Stacy Jo's bedroom, nothing to absorb the noise.

"We went to DOT when we heard they were going to start clearing and asked them if they could just skip the area in front of our house until the drill rig was gone. We knew it was a losing battle but wanted them to just wait until after the rig was gone. But they said they couldn't do that, so all those trees came down," said Stacy Jo.

Friends offered a positive spin: the Correias' expanded view of the inlet.

"But the thing is, if I had wanted to have those trees cut down I'd have married a woodcutter and we could have cut them down," said Stacy Jo. "We see the water all summer when we fish. At home, I like the trees, the seclusion."

In December, when Alaska's long hours of summer daylight are replaced by equally long hours of darkness and when Alaskan fishing families shift to the stillness of winter after months of being frantically harnessed to salmon's dash from ocean to spawning grounds, Robert called Hilcorp to complain. The non-stop activity, noise and bright lights on the drill pad were exhausting, he told them. It would be for only a few months, he was assured.

"And I thought, whew, I can get through that," he said.

When that proved not to be the case, Robert called Hilcorp again, only to be told the situation had changed. When he found strangers using spray paint to make markings in the right of way on the Correias' side of the highway, Robert again contacted Hilcorp, only to learn of other changes to the company's plans. Angered by the seeming inability to get a straight answer, Robert wadded up and threw away his Hilcorp business card with contact information.

"Every time I called, I was told things had changed," said Robert. "Tell me a lie once and your word is no longer good with me."

Buried in the right of way along the Sterling Highway is a high-pressure natural gas pipeline that snakes its way from wells south of the Correias' property to users to the north, the more heavily populated Anchorage area some 200 miles away. Some residents along the pipeline's route have worked with the Kenai Peninsula Borough to form utility special assessment districts in order to access the gas. For most, however, doing so is too costly. Therefore the gas passes them by. Hilcorp representatives told Robert a transfer station was needed to link activity on the Bartolowits' well pad to the pipeline. The site selected for the station was in the right of way, in the middle of the only access linking the Correias' property to the Sterling Highway.

"I said. 'This is my driveway. I put this driveway in. You're going to use my driveway?' They said, 'It's either that or we have to go down there and get stuck in the mud.' I said, 'Look, I don't want to be a jerk, but if you were in my shoes, what would you want to do? Would you want to live here?' They said no, they felt sorry for me, but it was progress," said Robert, recalling the conversation with disgust. And yet, he found himself sympathetic to the workers. "They're just people. Working for Hilcorp."

Another day, Robert saw a stranger parked on his property.

"I walked over and said, 'Excuse me, but this is my property.'"

On that occasion, Robert's politeness wasn't repaid in kind. The stranger's rudeness embarrassed even the crew of workers he had with him.

"I could see the look on their faces—they felt terrible because he was a jerk. They were just simple people working, not trying to trespass, but he was being a total jerk to me," said Robert, who turned to his faith to suppress an anger-fueled reaction. "I thought, 'I am a representative of Jesus Christ. I can't be a jerk.'"

Considering what might have happened had he followed through with his inclination to respond physically, Robert said, "I'm not a big man, so it's very fortunate I heard the spirit telling me to walk away."

A month later, equipment unexpectedly began rolling up the Correias' driveway to the pad made for the transfer site. When a couple of the trucks got stuck and tore up the road, Robert asked if someone was going to fix it. Yes, he was told. And they did, even following Robert's suggestions as to how it should be repaired.

"For three days, they had a whole crew of people. A roller, dump trucks. They flattened everything, hauled in recycled asphalt and did a really nice job. When they were done, I was amazed," he said. "I asked how much it cost and the guy said he thought they'd spent about $24,000."

Those repairs turned out to be the company's claim to the driveway. Or so it appeared to Robert.

"Almost every day one of their trucks would drive right up in my driveway and go in and out, just like it was theirs," he recalled. "It was driving me nuts."

Concerned about the level of activity at all hours of the day and night and the possibility of Sterling Highway motorists being blinded by the drill pad's bright lights, Robert contacted law enforcement in an attempt to prevent vehicle accidents. He also took his concerns to his state legislators and his representative on the Kenai Peninsula Borough Assembly. Nothing changed.

After receiving in the mail a card of pipeline safety information distributed by Hilcorp, the Correias' concerns multiplied. According to the card, copies had been sent to every residence or business identified as being near an underground natural gas or crude oil pipeline owned and operated by the company.

"At Hilcorp Alaska, LLC it is a priority to protect the health and safety of people and the environment and to conduct our operations responsibly," read the card's opening paragraph.

It included illustrations of pipeline markers and stressed the importance of noting the messages on the markers. It identified sights, sounds and smells that would indicate a leak and what to do if a leak occurred, and listed phone numbers to call immediately. It also included a "how are we doing?" card that recipients were asked to complete and return to Hilcorp.

"If these are possibilities, shouldn't there be something besides a brochure in the mail?" Robert wondered. When he asked neighbors if they had received a similar card, he learned he was the only one in the area who had. His conclusion was that Hilcorp "must expect it'll happen to me, but they don't have the guys come and talk to me."

As they became more and more uneasy about what was happening to the neighborhood, Robert and Stacy Jo's sense of desperation began to grow. They weren't in a financial position to buy another place, but perhaps there was another way. Robert approached Hilcorp about the possibility of the company buying the Correias' property. It was "very possible," he was told. Someone would get back to him. When he didn't hear anything, Robert placed another call, only to be told Hilcorp's purchase of his land was not possible after all.

On another occasion, the Correia family was asleep. A few hours away was a 5 a.m. get-up and then the long drive to Anchorage for a daughter's medical appointment. Shortly after midnight, the beeping of a front-end loader backing up and the clanking of pipe being loaded onto a trailer awakened them. Robert's patience was gone. He put on his shoes and robe, grabbed a flashlight and went across the road.

"I said I live right across the highway and could they please not do this in the middle of the night. 'Oh yeah' they told me, 'we work round the clock and we can't turn off the back-up beeper. Sorry. But we're almost done here and then we'll be gone for a couple of hours and back around 3 to do some more,'" Robert said of his exchange with the equipment operators.

Angered by the lack of consideration, he found what he sleepily thought was the business card with Hilcorp contact information. Forgetting he'd already thrown it away, he called a number written on the back that he hoped was a Hilcorp representative's home phone. When a woman answered, Robert launched into a heated and detailed account of his frustrations.

His tirade was met with silence. When Robert asked if the woman on the other end of the phone was still there, he was chagrined to be told the person he'd called in the middle of the night wasn't a Hilcorp employee. It was the mother of a friend who happened to be a Realtor.

. .

IN THE FALL OF 2014, after more than a year of dealing with the round-the-clock activity on the nearby drill pad and its devastating effect on their lives, Robert and Stacy Jo made the difficult decision to give up the fight.

"There's so many beautiful places in Alaska, but I can't think of any place less beautiful than where we live," said Robert of what had happened to the sanctuary he and his wife had created for their family.

The Correias put their property up for sale and their belongings in storage, found places for their animals, and moved themselves and their six children into the bed and breakfast for the winter.

"Robert can't sleep at night. All night, every night he's pacing. He just can't stand it. I've tried thinking he'll get used to it, but no, he hasn't gotten used to it," Stacy Jo told me the evening I met with them at the B&B.

Once the decision to sell had been made, the couple settled on a price that equaled the cost of the house and original piece of property, "basically giving away the second piece," said Robert. A couple who worked at two different locations on the Kenai Peninsula were interested—the Correias' place offered a central location. Because the buyers had four horses the barn and corral were an added inducement.

"Who in their right mind would want to live next to that stuff? The only reason is that you don't have any money or because of your job," said Robert.

"It's highly different for us because we lived there when it was quiet. When there were trees. When there was no one across the highway. For someone else to come along, even if it was us, we'd think it was a steal," he said. "Yeah, there's a drill rig there and there's no trees, but trees can grow back. The only thing is we know better. We were there for seven years. Six years when it was pristine."

"When I'd drive in the driveway, into our world, it felt like an enchanted place," Stacy Jo reflected. "Then all of a sudden it was naked to the highway, Halogen lights shining at us, drilling going on, helicopters. It felt like an alien invasion."

"Every day when I'm moving stuff," Robert continued, "I get up in the morning with my stomach in knots. I go over there and look out the windows on one side and it's beautiful. I see the barn, the horses, so much work I've done, and it's beautiful. Awesome. What am I doing? It tears me up inside.Then they start the rig up and I think, 'Thank you, Lord, for reminding me.'"

The money from the sale of their place will help the Correias "start from scratch somewhere else," said Robert. In a perfect world, that "somewhere else" will be miles away from where Hilcorp or any other oil or gas company decides to operate.

"Couldn't they set aside part of Alaska as a space where people can live and not be bothered by oil companies?" Robert asked.

As they try to put the events of the past two years behind them, Robert has one wish for Hilcorp.

"I hope they're blessed with the understanding that we're put on this earth to treat each other like we love each other," he said. "The best thing I can do is treat them with the love and respect Christ gives me."

But he is well aware of his own shortcomings, he says, the full measure of anger that knots his stomach and the temptation to lash out at its cause.

"The truth of the matter is that almost every day I think about the things I want to do, that I might do, things that are so crazy. That's one reason I need to leave from there," he said.

In the midst of a temporary situation, with the family's past a closing chapter and their future uncertain, Stacy Jo's guiding star was a desire to distance her husband, herself and her family from the inner turmoil caused by circumstances not of their making.

"It's the only reason," she said.

KATIE

"How do you prove it?"

BEING WELCOMED TO KATIE KENNEDY'S HOME has been like inhaling Alaska. In one big gulp. A force field of energy surges through and around her like the tide-driven waves pulsing along Alaska's thousands of miles of shoreline. Her welcoming hug sparks memories of homecomings after long journeys. Her blunt honesty has an in-your-face quality—the earth-shaking slip of tectonic plates comes to mind. Katie's love affair with this part of the world is mirrored by the books on her shelves, the artwork on her walls, the beach rocks surrounding her woodburning stove.

Born in Montana, Katie came to Alaska with her parents and siblings in the mid-1960s. Her father, the late Michael Kennedy, was the first director of the Museum of History and Art in Anchorage. He went on to become director of the Alaska State Museum in Juneau and, later, the state historian. His contributions to the state and his efforts "to augment the quality of life for all his fellow men and women" were recognized by the Alaska Legislature "In Memoriam," Jan. 29, 1993.

"Every time you see what's historic that's been saved, it's my dad's doing. He did so much for this state," Katie told me proudly as we sat in her sun-warmed home, its many windows offering views of Cook Inlet and the distant Aleutian Range.

Raised in an atmosphere of artifacts from Alaska's past, Katie inhaled the state's history and exhaled it through her views and values. Her immersion in the commercial fishing industry, an occupation dependent on tides and currents and the migratory calendar of fish, served to deepen that love of her adopted homeland.

"I was always fishing. Commercial fishing on a boat. As a crew member. Herring fishing in Togiak and Prince William Sound. Longlining on a halibut schooner. King crab fishing out west. I've been on a seine boat, gillnetted and setnetted," she said.

When the *Exxon Valdez* plowed into Bligh Reef in Prince William Sound in the early morning hours of March 24, 1989, Katie was living in Homer. News spread across Alaska, the nation, and the world that the tanker's ripped hull was spilling millions of gallons of crude oil into Alaskan waters, fouling more than a thousand miles of beaches.[1]

"I was devastated," she said. "It just got worse and worse and worse."

As oil company representatives struggled to contain the damage, people from all walks of life scrubbed shorelines, gathered oil-soaked carcasses of creatures whose lives were the currency the tragedy exacted and fought to save those still breathing. Katie was among the responders. She worked long hours with a sea otter treatment effort in Seward and tended to seabirds struggling for breath at a Homer recovery site. With every minute waged in the battle to save animals' lives, her resolve deepened to separate herself from the industry behind the tragedy.

"I worked all summer and never took one dime," Katie told me. "Not one dime. It was blood money."

In 1990, at Concert on the Lawn, an annual music festival and fundraiser for KBBI, Homer's public radio station, Katie met Don Erwin. The couple fell in love with a four-acre parcel of land on the bluff some 40 miles north of Homer, near Ninilchik, and established that as their home base. Don operated a sportfishing charter business in the summer and turned the structure already existing on their newly purchased land into a home large enough for Katie to establish a bed and breakfast. Its bluff location, view-filled windows and tall, two-story walls were the perfect venue for Katie to express and share with others her strong sense of place.

Shortly after Don and Katie bought the property, a couple they didn't know drove into their yard.

"They had come from Fairbanks and were just driving by and said, 'Do you know of any good property?' And I said, 'There's the most beautiful piece right down there,'" Katie said, recalling the day she directed them toward a nearby 30-some-acre parcel the couple later purchased.

Eventually that property, or more accurately the potential oil and gas deposits beneath the surface of that property, drew the attention of Marathon Oil. In 2012 Hilcorp Energy Co., a privately held Texas-based firm, acquired Marathon's Cook Inlet assets, including the Susan B. Dionne pad—named after the woman who with her husband had asked Katie and Don about property in the area and located on the parcel's bluff boundary. Activity on the pad lit up the night sky, making it clearly visible from Katie and Don's home.

As Marathon developed its activity at the site, company representatives approached Katie with lease offers that would allow them access to what might be beneath the surface of her land. She repeatedly refused the offers. Should such a request come in Katie's absence, as it did one afternoon when her sister happened to be alone at Katie and Don's home, the answer was the same.

"One drove in with his Texas accent. 'Is Miz. Kathleen Kennedy here?'" Katie said, mimicking the southwestern twang of many in the oil industry. "My sister said, 'I can tell you're someone she'd just tell to fuck off' and he got in his truck and left."

"That's how I knew I had a lot of oil underneath me, because they were hitting on me," she said. But this wasn't the first clue Katie had that there was oil on her land.

"Years ago when I had my water tested, they found out it was a pure artesian well and the guy at the time told me that was a significant indication of huge reserves of oil and gas. Who'd have thought?" she said, laughing at the irony of living atop a resource representing an industry she detests and distrusts. Then in 2003 she allowed a company named Veritas to conduct seismic surveying on her land.

"At that time it was revealed to me that there were huge reserves," Katie said. "We're talking oil, not just gas."

When activities at the drill pad passed from Marathon to Hilcorp, the new operator also approached Katie about a lease to grant Hilcorp subsurface access to her land. With the continued interest and her concerns mounting about possible negative impacts that drilling operations might have on Don's and her home, their water source, septic system and stability of the nearby bluff, Katie sought legal advice.

"I got the attorney so I could find out what's going on," she said. "We don't even have community meetings any more to tell anybody. Yes, they did have two, but every goddamn thing they said turned out to be a lie. What I think is that they honestly believe people are too afraid of their power to stand up to them."

The attorney's advice was something Katie didn't expect.

"He suggested I sign a lease," she said. With a signed lease spelling out the work that would be done on her property, Katie was told, she had a better chance of holding the operator liable should any damage occur.

Then came the news that the boundaries of the unit had been redrawn, placing Katie's property outside the Susan B. Dionne production site. According to the new boundaries, Hilcorp would not be producing from any pool of oil or natural gas beneath the surface of Katie's land. That struck her as highly suspicious. It didn't fit with the interest expressed by more than one operator in gaining subsurface access, but it fit perfectly with her already existing distrust of the oil and gas industry.

"What I've been told is that if an oil company took one drop from underneath me without paying me a royalty, it was big. But how do you prove it?" she asked.

. .

SUCH A SITUATION is exactly what landed Buccaneer Energy in court after Cook Inlet Region Inc. (CIRI), an Alaska Native corporation, proved partial ownership of a natural gas field in Kenai. In 2012, Buccaneer began producing from two wells on property owned by the Alaska Mental Health Land Trust. CIRI suspected that a portion of gas produced from the wells was being drained from the corporation's adjacent parcel. The corporation filed suit against Buccaneer in state court and through the Alaska Oil and Gas Conservation Commission pursued payment of royalties CIRI believed were owed.

"There were geologists brought in, ours and Buccaneer had theirs, to determine what the geologic features of the gas pool were, and it was determined that a certain portion of the gas resource was from CIRI land," Jason Moore, CIRI's corporate communications director, told me over the phone. "It's my understanding that even Buccaneer's geologists believed it was likely they were pulling resource from CIRI land."

Buccaneer filed for bankruptcy in 2014, with an outstanding debt of more than $2.1 million to unsecured Alaska creditors. The company's assets, including the Kenai field and its associated leases, equipment, data, permits, office equipment, furniture and software, as well as offshore and onshore Cook Inlet tracts and 13 contracts, were won in auction by AIX Energy LLC for $44 million. AIX is a subsidiary of Meridian Capital International Fund, which financed a portion of Buccaneer's work in Cook Inlet.[2]

In January 2015, a Southern Texas bankruptcy court approved a settlement between CIRI and Buccaneer releasing the two parties from possible liabilities and removing CIRI's $5.75 million proofs of claim against Buccaneer.[3] CIRI then entered into an agreement with AIX, which "paid the back royalties for that gas," said Moore, declining to disclose the exact amount paid.

Don has advised Hilcorp of his and Katie's suspicions about their land, but, Katie said, "they keep telling us, every time Don calls them, 'We're nowhere near you.'"

A corporation calling a company like Buccaneer to question is one thing, Katie said, referring to CIRI's experience, but "Katie going up against these people? How do you prove it?"

· ·

OFFSHORE SEISMIC testing in Cook Inlet also worries Katie. For decades, sound waves have been used on- and offshore by the petroleum industry to identify oil and natural gas reserves. The resulting data also were used by U.S. Geological Survey scientist Peter Haeussler and others to map a network of folds and faults beneath Cook Inlet that have the potential of creating earthquakes with a magnitude of 6.0 or greater. The scientists' findings resulted in a warning to oil and gas producers in the Cook Inlet area.

"I think the oil companies should assess whether pipelines can be compressed as the faults shift. The faults could produce earthquakes large enough to rupture pipelines," Haeussler has said.[4]

Because of the potential impact on sea life, seismic testing falls within the permit and authorization requirements of the Endangered Species Act (ESA) and the Marine Mammal Protection Act

(MMPA), when such activity could lead to the "taking" of a protected species. "Take" is defined by the ESA as harassing, harming, pursuing, hunting, shooting, wounding, killing, trapping, capturing or collecting or attempting to engage in any such conduct; it is defined by the MMPA as harassing, hunting, capturing, collecting or killing or attempting to harass, hunt, capture, collect or kill any marine mammal.

For its plans to operate a drilling platform in the lower Cook Inlet, BlueCrest Energy applied to the National Marine Fisheries Service (NMFS) for an Incidental Harassment Authorization. For the multiple activities related to the operation, "the primary impact of concern is the effect the noise generated by these operations could have on local marine mammals," the July 2014 application read. Species within the operating area included gray, killer and minke whales, harbor and Dall's porpoises, and harbor seals.[5]

Apache Alaska Corp. petitioned the NMFS in 2014 to issue regulations for the "non-lethal unintentional taking of small numbers of beluga whale and other marine mammals incidental to oil and gas exploration seismic operations and all associated activities in Cook Inlet for the period of five years beginning March 1, 2015, extending through February 29, 2020." The petition asked that the regulations identify "permissible methods of non-lethal take, measures to ensure the least practicable adverse impact on these species; and requirements for monitoring and reporting."[6]

But that's whales, seals and porpoises. Katie's concerns focus on a possible link between seismic testing and the decreased number of razor clams on Ninilchik and neighboring beaches.

For generations these sharp-edged bivalves have been a valued food source for Ninilchik-area residents. During dark winter months, when fresh meat was at a premium in the once-isolated village and low tides were at their most extreme, founding families used lanterns to locate dimpled sand signaling a clam just beneath

the surface. Once such was spotted, an individual would frantically dig through the grit, elbow deep in icy water, eager to clasp
the clam's shell before its strong digging foot secured its escape.
Visions of a pot of chowder simmering on the stove, a platter
heaped with tasty fried strips or servings of juicy fritters for the
family made the effort worthwhile.

As Alaska's population grew, so did the popularity of year-
round clam digging. The Alaska Department of Fish and Game
reports that in 1969, clam diggers made 8,600 visits to Cook Inlet
beaches and took home 279,500 razor clams. The annual harvest
by the mid-1980s was about one million razor clams. In 2006,
from the Kenai River to the tip of the Homer Spit, a distance of
about 90 miles, the allowable limit of razor clams that could be
harvested in a day was 60 per person. In 2013, that was reduced
by more than half, to 25 per day. In 2014, Fish and Game shut
down clam harvesting along Ninilchik beach. A year later, the closure was extended the full 90 or so miles, from the Kenai River to
the tip of the Spit.

That reduction in harvest limits and the subsequent closures
followed an unusual event in November 2010, when thousands of
razor clams, most of them two years old, washed up along an
eight-mile stretch of Ninilchik beach. In just one 70-by-30-yard
patch, Alaska Department of Fish and Game biologists estimated
2,700 clams covered the gravel.[7]

One of those biologists, Nicky Szarzi, attributed the phenomenon to a strong wind and large tides combined with an unusually
high number of 2-year-old clams. Locals, many of them descendants of Ninilchik's original families, argued that this wasn't the
first time Ninilchik had been hit with large tides and strong winds
but it was the first time they'd seen clams lining the shore.

"We've had bigger storms than this," Edna Steik said at the
time. "This is weird." Edna's great-great-grandparents, Grigorii
Kvasnikoff and Mavra Rastorguev, and their children were among

the first families to settle in Ninilchik in 1847. Edna died at age 77 in 2012, in the village in which she was born.

Some attributed the phenomenon to warmer winters. Others, like the late Frank Mullen, a commercial fisherman who lived on the peninsula for more than 60 years, blamed the state's poor management practices.

"The base population of clams was overharvested to such an extent that the density of clams necessary for spawning and future recruitment was reduced to a point of vulnerability," Mullen wrote in a March 12, 2015, *Homer News* "Point of View." Of the 2015 closure, he said, "This should have occurred years ago. I sincerely hope that it is not 'too little too late.' If the fishery is reopened in the future, the bag limits should be dramatically reduced."[8]

For Katie, though, the bottom line is simple.

"The clams are gone. I really think it's this seismic stuff," she said.

Although multiple studies about the impact of seismic activity on marine mammals have been conducted, research specific to razor clams is sparse. In a section dealing with razor clams' need for relatively high levels of dissolved oxygen, part of a 1989 report on Pacific razor clams for the U.S. Fish and Wildlife Service and U.S. Army Corps of Engineers, the authors wrote, "In an era of increasing nearshore oil exploration, in the event of an oil spill, subsurface oxygen may be affected." However, they added, "We are not prepared to say how that would impact razor clams."[9] Contacted by email in 2015, one of the report's authors, Dennis Lassuy, said he was not aware of any subsequent studies on razor clams.

Reporting for the *Homer News* on the 2014 closure of Ninilchik beaches to clamming, I asked Carol Kerkvliet, a state fishery biologist, if the impact of seismic testing on razor clams had been considered as a reason for the decline in the clam population; she said no. Then, in 2015, marine educator Carmen Fields shared with me an article found by Kerkvliet that had been published by the University of St. Andrews in Scotland on the findings of researchers from the Scottish Oceans Institute at St. Andrews, the University of La Laguna, Canary Islands, and the University of Auckland, New Zealand.[10] It indicated that marine invertebrates, including shellfish, "suffered significant body malformations after being exposed to noise." In a sound playback experiment on New Zealand scallop larvae, 46 percent of the exposed scallops had developed body abnormalities.

"The strong impacts observed in the experiment suggest that abnormalities and growth delays could also occur at lower noise levels in the wild, suggesting routine underwater sounds from oil exploration and construction could affect the survival of wild scallops," the article reported.

It went on to quote team leader Dr. Aguilar de Soto of the University of St. Andrews and the University of La Laguna: "Fishermen worldwide complain about reductions in captures following seismic surveys used for oil explorations. Our results suggest that noise could be one factor explaining delayed effects on stocks."

The article closed with a plea from senior research fellow Dr. Mark Johnson of St. Andrews: "Between shipping, construction and oil explorations, we are making more and more noise in the oceans. . . . It is important to find out what noise levels are safe for shellfish to help reduce our impact on these key links to the food chain."

· ·

KATIE'S FRUSTRATIONS with the oil and gas industry have grown as every concern she's voiced and every question she's asked have been met with what she believes is inaccurate information.

"When I told them I'd been reading about seismic work and new techniques that were more sensitive to the environment, they said, 'We're doing all new technology.' When I said 'You're certainly going to be putting locals to work' they said, 'Oh yes, we're getting all locals,'" said Katie, who has seen crews she knows aren't from the area. "I gave them the benefit of the doubt and my friend said, 'Oh, Katie, you're such a wimp. They told you everything you wanted to hear.'"

Those frustrations have added to others as over the years Katie has witnessed changes taking place in Alaska that are tied, one way or another, to the state's reliance on oil and gas revenues.

"Where I first really felt upset was when we quit giving away the longevity bonus to our elders," Katie said of a state program that began in Alaska in 1972. The Senior Citizen Longevity Bonus Program gave the state's residents who were 65 and older and who had been in Alaska since statehood monthly checks of as much as $250. In 1984, the Alaska Supreme Court ruled the program unconstitutional and changed the eligibility requirements. In 2003, with the state budget threatened by a drop in oil revenues, then-Gov. Frank Murkowski, father of Lisa Murkowski, Alaska's current senior senator, brought the program to an end when he vetoed funding for the $44 million program.[11]

"I had to watch my two favorite neighbors leave Alaska because that $500 made all the difference to them. They couldn't live here anymore," said Katie.

The demise of Alaska's Coastal Zone Management Program stood out to Katie as yet another area where the oil and gas industry had left its mark. Congress passed the Coastal Zone Management Act in 1972, as a partnership between the federal government and coastal and Great Lakes states and territories. It is administered by the National Oceanic and Atmospheric Administration, its goal to "preserve, protect, develop, and where possible, to restore or enhance the resources of the nation's coastal zone." Its flexibility allows program partners to address specific challenges while working within state and local laws and regulations. All 35 coastal and Great Lakes states and territories participate in the program.[12] Or did until 2011, when Alaska, the state with the most coastline, withdrew from the program.

During the 2010–2011 legislative session, with a lack of consensus in the state Senate and House regarding continuation of the Coastal Zone Management Program, then Gov. Sean Parnell let the program come to an end. Almost immediately, the Alaska Sea Party formed to lead a grassroots effort to recreate the program. In 2012, more than 33,000 Alaskans signed a petition to put Coastal Zone Management on a statewide ballot. A series of 10 hearings on the ballot initiative were held around the state, with then Lt. Gov. Mead Treadwell serving as moderator.

Resource developers, such as Lorna Shaw, external affairs manager for Sumitomo Metal Mining Pogo LLC, spoke against the ballot measure at the Kenai hearing, saying it would "put the brakes on investment and projects because it adds more red tape." Lisa Parker, a spokesperson for Cook Inlet developer Apache, said the program would stifle Apache's ability to move forward with projects.[13]

Kodiak Borough Mayor Jerome Selby, a sponsor of the initiative, charged the opposition with "boogeyman fear-mongering, plain and simple."[14]

In the state's 2012 primary, the initiative was defeated, with 76,440 voters opposing renewal of the Coastal Zone Management Program and 46,678 votes supporting it.[15]

"That program was the only power communities had when it came to development in their areas," said Katie.

Another disappointment for her followed Peter Micciche's election to the Legislature in 2012. In addition to being a commercial fisherman and the former mayor of Soldotna, Micciche is superintendent of the ConocoPhillips liquefied natural gas plant in Kenai.[16] His legislative assignments have included serving as co-chair of the Special Committee on TAPS (trans-Alaska pipeline system) throughput—a committee formed to reverse or reduce the historical decline in the quantity of oil being shipped through the pipeline system; as a member of the Special Committee on In-State Energy, during the 2013–2014 session, and as vice-chair of the Finance Committee and co-chair of the Special Committee on Energy in 2015.

When Senate Bill 21, sponsored by then-Gov. Parnell, was introduced to restructure Alaska's oil and gas production tax, Micciche and Sen. Kevin Meyer, also a ConocoPhillips employee, declared "perceived" conflicts of interest and asked to be recused from the Senate vote. Other senators objected to their request and, by existing rules, both Micciche and Meyer were required to vote. The bill passed the Senate 11–9.[17]

Questioned about the "perceived" conflict of interest, Micciche responded, "The press, in my view, irresponsibly doesn't take the time to get to know you, doesn't look at your record—30-year record in the community—and they almost make it sound like you are this sort of new corporate plant that showed up one day to do the work of the corporation instead of who you really are."[18]

Following passage of SB 21, a "Vote Yes! Repeal Giveaway" nonpartisan group was formed to urge repeal of the legislation, claiming it gave unfair tax breaks to the oil industry, especially the three biggest producers, BP, ExxonMobil Corp. and ConocoPhillips. The group was led by Vic Fischer, a delegate to the Alaska Constitutional Convention in 1955–1956 and a former territorial and state legislator. At a rally held in Homer, one of many around the state, Fischer recited a portion of the state's constitution that provides "for the utilization, development, and conservation of all natural resources belonging to the State including land and waters, for the maximum benefit of its people."

"SB 21 was not for the maximum benefit of the people," he told the crowd.[19]

In spite of the group's efforts, however, the repeal lost in the 2014 primary, 89,608 voters in support and 99,855 voters opposing it.[20]

In Katie's view, "The corrupt corporate cronyism, the fact that we can have someone work for ConocoPhillips and vote on something as crucial as giving that type of tax incentive to the three largest oil companies in the world just took us down the tubes."

In a state where residents boast about their independence and sport bumper stickers that declare "We don't give a damn how they do it Outside" (Outside being a reference to any place outside Alaska), Katie contends a crucial element is lacking: "What it all boils down to is that there's just no common sense."

For more than 60 years, Katie said, she has campaigned for causes she holds dear and faced criticism from those who think differently. She has attended meetings to gather information and raised her hand to ask questions so she could increase her understanding. She also, she said, has practiced what she preached.

She was a leader in high school supporting the original Earth Day, conceived by Gaylord Nelson in April 1970. "I've always been an active 'greenie,'" she said, "practicing recycling and conservation of resources long before it was mainstream."

Her desire "to keep the planet beautiful for all to enjoy" and a belief that oil is earth's shock-absorber has spurred her belief that extracting it in greater and greater amounts, rather than broadening the use of alternate energies such as wind, solar and geothermal, is "extremely short-sighted."

"Therefore, I thought it was great irony when I bought my beautiful property on the bluff in Ninilchik and found out that I have the mineral rights," she said.

That sense of irony turned to bitterness with the increasing pressure from first Marathon and then Hilcorp for her to sign a lease, only to discover that the boundaries of the unit had changed, a move Katie suspects was made to keep the producers from having to compensate those who owned mineral rights.

In the summer of 2015, Katie took steps to sell the four acres that have been Don's and her home for the past two decades. With oil and natural gas development increasing to the north and south, as well as offshore, she's convinced it's time to move on. Her plan is to pursue interests in geothermal energy and agriculture in Oregon.

"I have no desire to live in Alaska anymore," Katie said. "Oil and gas have drawn a different type of people up here to live than the people who were here when I was growing up. Not the very self-reliant and independent people who were helpful to neighbors and communities. ... I can't bear to watch this anymore."

The new owners of the four-acre bluff property will get the house Katie and Don built and in which they have introduced many visitors to the Alaska they love. They will get inlet winds on which bald eagles glide just beyond the bluff. They will get the sound of the surf. They will enjoy sunsets over the Aleutian Range at the end of the day and the aurora dancing overhead on clear winter nights.

What the sale does not include is the one piece Katie has vowed to safeguard.

"I'm keeping the mineral rights," she said.

Darwin and Kaye

"I want to be in Alaska for as long as I can"

D ARWIN AND KAYE WALDSMITH CAME TO ALASKA at different times, from different places. Separately, they came to teach. What they found, besides each other, was a people and lifestyle they valued. Half a century later, the Waldsmiths are doing their part to protect their adopted home of Ninilchik from a presence seemingly determined to take over the neighborhood.

Darwin was headed to Australia in 1966. He planned a one-year stop in Alaska to teach before continuing from his home state of Wisconsin, by way of Illinois, to the southern hemisphere. The idea of leaving Alaska disappeared once he began introducing himself.

"I'd never realized people could be that caring," he said. "The different places I'd been were large population areas, considerably larger than Ninilchik, and kids had a lot of things they could do to stray—but here, everybody kind of took care of everybody."

When Darwin lists his first Ninilchik acquaintances, the names are a who's who of the village's founding families, several generations removed.

"I always considered Martha Jensen as my Alaska mother," he said, recalling the relationship that developed with Martha Kelly Jensen as well as with her husband, Torvald Jensen. Martha's great-grandparents, Mavra Rastorguev and Grigorii Kvasnikoff, headed one of the five families considered the nucleus of Ninilchik's first permanent settlers in the mid-1800s.[1]

His days at Ninilchik School, a first- through twelfth-grade facility with about 100 students at the time, were every bit as busy as a teacher in that size school would expect. Darwin taught woodworking, geometry, seventh- and eighth-grade industrial arts and mechanical drawing.

With summers free, he made what he calls a "pivotal step" when he became involved in Alaska's multifaceted commercial fishing industry. He fell in love with the intense energy of the fishing seasons, the adventure of being on the water and the opportunity to be in the outdoors. He gillnetted on Cook Inlet and Prince William Sound and in Southeast Alaska, seined on Cook Inlet and Prince William Sound, fished for halibut on Cook Inlet and around Kodiak and for king crab in Kachemak Bay—"kindergarten crabbing," he called it, laughing, in comparison with the infamously brutal conditions under which crabbers operate in Alaska's Bering Sea.

· ·

KAYE, A NATIVE New Yorker, migrated north in 1967 to teach in Anchorage, Alaska's largest city. The city setting made her introduction to Alaska markedly different from Darwin's, but it was equally memorable for the change of pace from her home state and for the people she met. Her initial circle of friends hailed as she had from points outside the state.

"People came from all over the Lower 48, and if you were sitting around a campfire there were always lots of stories," she said.

The exception to the new friendships Kaye forged was one person she met at a dinner with neighbors shortly after her arrival. The individual made an uncomfortable impression she never quite forgot. Then, in 1983, Kaye's memories of Robert Christian Hansen surfaced when Hansen, who became known as "Butcher Baker," was arrested and convicted of more than a dozen abductions, rapes and murders of women in the Anchorage area between 1971 and 1983.

After Kaye's introduction to Alaska in Anchorage, she accepted a teaching assignment in Fairbanks, the state's second largest city. In 1971, at a summer course on environmental education in Seward, on the east coast of the Kenai Peninsula, she met Darwin, who also was enrolled in the class. Six months later they were married, and Kaye moved to Ninilchik. When the 1972–1973 school year got under way, she filled in as a substitute teacher and tutor at Ninilchik School and finally landed a full-time teaching position in 1978.

Darwin continued to teach until 1987. Fishing has been harder to give up, however. As of the 2015 season, he was still setting his summer clock by the opening and closing of the state's various fisheries. Kaye taught school until 1997.

· ·

THREE YEARS after Darwin and Kaye married, they bought a 50-acre piece of property on the inlet side of the Sterling Highway and built a home just three miles from Ninilchik School. Tucked into a grove of birch trees, the two-story house has a beautiful view of Cook Inlet and the range of mountains bordering the inlet's western shore. From the Waldsmiths' tree-framed vantage point, two volcanoes in a string of five stand tall on the horizon, Iliamna and Redoubt. It is not uncommon to see puffs of steam rising from their tops. Out of view are Douglas and Augustine to the south, Spurr to the north. Iliamna last erupted in 1953, before either

Darwin or Kaye had come to Alaska. Redoubt is more restless. Its most recent eruption was in 2009, when a billowing plume of ash spread across the inlet to the Kenai Peninsula and halted air traffic in and out of Ted Stevens International Airport in Anchorage.

In 1983 the Waldsmiths purchased an additional 70 acres adjacent to the parcel on which they had settled. The purchase agreements for both pieces of land granted them not only surface rights but also 75 percent ownership of subsurface rights.

Today, marking an entrance to the Waldsmiths' property from the Sterling Highway, and to neighboring driveways on either side, are orange "Warning: Natural Gas Pipeline" signs. Since 2003, a high-pressure line has transported natural gas along the right of way from wells on the southern peninsula to other Alaska markets. An even newer addition to the neighborhood is Hilcorp Alaska LLC. In 2011, the Texas-based independent acquired Union Oil Co. of California's Cook Inlet assets, and the following year the Cook Inlet assets of Marathon Oil Co. Hilcorp strengthened its Alaska presence in 2014 by also acquiring a large portion of BP Exploration (Alaska) Inc's North Slope assets.[2]

Hilcorp is operator of the Paxton Lateral pad half a mile from the Waldsmiths' home. Located on the edge of the bluff overlooking the beach, Cook Inlet, and the same mountains that can be seen from the Waldsmiths' living room, the pad is the southern base of operations for Hilcorp's Susan Dionne/Paxton Participating Area, a 2004 natural gas discovery that previously was operated by Marathon. The Susan Dionne/Paxton, Falls Creek and Grassim Oskolkoff participating areas lie within Hilcorp's 25,824-acre Ninilchik Unit. The unit began producing in 2003 and as of mid-2015 had eight drilling pads with plans for a ninth to be added.

The winter of 2014–2015 was a particularly busy one at Paxton Lateral. Lights from drilling operations illuminated the sky above nearby homes, including the Waldsmiths', and filled the air with noise. The Waldsmiths are not merely neighbors. Their entire acreage falls within Hilcorp's Ninilchik Unit boundary.

Despite the physical proximity, Darwin and Kaye have been determined not to become involved with the oil and gas activities around their property. Seismic crews have made continual attempts to pinpoint locations of possible gas and oil deposits, including beneath their house. Asked repeatedly to sign leases allowing access to their land, they have sought legal advice as the southern Kenai Peninsula's oil and gas industry grows around them. Mindful of development that has occurred in other areas of the state, the Waldsmiths have cautiously guarded their rights, both surface and subsurface.

"Oil development has taken some of the prettiest land there is. We just don't want to see that happen to our community," said Kaye.

Having heard pro-oil and gas arguments that champion the number of jobs the industry has created for Alaskans, Darwin believes that factor has to be weighed along with the positive and negative impacts of resource extraction. Like most Sterling Highway residents living along the natural gas pipeline that runs beneath their driveways, the Waldsmiths lack access to its contents. Were it made available, Darwin recognizes it could reduce their dependence on electricity and heating oil and result in a savings that would ease the cost of living where he and Kaye have chosen to reside.

"I want to be in Alaska for as long as I can," Darwin said. "Could that help us get our fuel bill down to half? It's not inconceivable."[3]

· ·

WITH OIL AND GAS exploration and production spreading around them, both on- and offshore, the Waldsmiths do their homework, trying to grasp the overall picture: the industry's plans, state laws pertaining to oil and gas, and the rights of homeowners. Keeping up with industry activity is no small task. In the Wald-smiths' area alone, Hilcorp's plans for 2014 initially called for six exploratory wells in the Ninilchik Unit. Toward year's end, the work had expanded to 11 wells, including an 11,000-foot well on the Paxton Pad and two wells slightly north, between the Paxton and Susan Dionne pads.

A few miles south of the Waldsmiths, Hilcorp has been explor-ing its 22,657-acre Deep Creek Unit, originally operated by Stan-dard Oil in 1958, then Unocal and now Hilcorp. After taking it over, Hilcorp drilled three wells and planned to expand.[4]

Not surprising, given the level of activity, Darwin talked of the benefit of a group of neighbors meeting to share what they know and help each other understand oil and gas development not just in the area but around the state and nation. He offered to forward several documents he had been reading that although "dry" might be helpful in my research. It was clear that Darwin's interest sur-passed knowing where drilling is occurring and rates of produc-tion—true to his word, within hours after I returned home the documents arrived in my email inbox, among them technically de-tailed descriptions of directional and horizontal drilling operations. There were links to documents on forced or compulsory pooling, the right granted by a state regulatory body for a company to in-clude adjacent tracts in its drilling unit. Darwin also included in-formation on unitization, the grouping of wells in an area, and provided copies of laws pertaining to different facets of oil and gas activity in Alaska and other locations in the United States.

Asked what could be done to foster positive relations between residents and oil and gas developers, Darwin offered several sug-gestions. For starters, he said, property owners need to know their rights. Referring to the Homestead Act, he expressed concern for "the stripping away of landowners' rights."[5]

If property rights are not being honored "and the state isn't looking out for you," Darwin said, "it would have to be the federal government making it work. I've sat here and mulled it over and that's what I've come up with. That would be my first avenue." An advocate would be required, a public defender of some sort, who would "at least listen to you and help you get your rights enforced."

Another of Darwin's suggestions rests on voters' shoulders. It requires careful screening and keeping out of office candidates with ties to oil companies.

On the topic of alternative energy, Darwin sees several challenges. Nuclear power "has some serious drawbacks." Solar is problematic in an area that doesn't see much sun during the winter months. Winds are unpredictable. Tidal power, on the other hand, offers some hope for Alaskans. Technology to harness Cook Inlet's strong diurnal tides might take a decade to develop, "but even if it didn't work, at least we'd have tried," Darwin said.

A renewable source of energy, tidal energy relies on the powerful surge of rising and falling ocean tides to generate electricity. The Sihwa Lake Tidal Power Station in South Korea opened in 2011 and, with an output of 254 megawatts, is the largest tidal power plant in the world; second is La Rance Tidal Power Plant in France, in operation since 1966 and with an output of 240 MW. Other large facilities are in operation in Scotland and Canada.[6] The highest tides in the United States can be found not far from Ninilchik, at the head of Cook Inlet, near Anchorage, where the difference between highs and lows can exceed 40 feet.[7] Homer Electric Association, electricity provider for the Ninilchik area, is currently working with Ocean Renewable Power Co. to explore the possibility of putting the inlet's tidal energy to use.

· ·

IN ALASKA, WHERE more than half of the total budget and 90 percent of its general fund come from oil revenue, the ups and downs of the cost of a barrel of oil can have a huge impact.[8] In a year like 2015, with oil prices falling and Gov. Bill Walker and the

Legislature straining to make ends meet, Darwin had one word for Alaska's future: bleak.

Throughout the summer of 2015, Walker and Lt. Gov. Byron Mallott hosted a series of conversations around the state that focused on closing the gap between what it costs to run the state and the lack of oil dollars to make it possible.

"As most of you know, the price of oil fell precipitously over the past year, and with it, Alaska's biggest source of revenue," Walker wrote in an opinion piece for the *Alaska Dispatch News*.[9] "Alaska's budget for the coming year proposes to spend $5 billion in state general funds. We expect to receive only $2 billion in general fund revenue. That leaves a gap of $3 billion. When your family's income plummets, you have two basic options: spend less, and earn more. The State of Alaska is no different."

Saying it was impossible to "cut our way to a balanced budget," Walker called on Alaskans to help create "a sustainable future for ourselves. . . . With oil revenue alone unable to sustain us, how do we chart a new course? It's time for us to pool our collective energy and ingenuity, and take control of our destiny."

In December 2015, Walker unveiled the "Walker-Mallott Administration's New Sustainable Alaska Plan." It addressed the state's budget deficit with a combination of spending cuts, new revenue, wealth management and investment.[10] The proposal included putting oil tax revenue into the Alaska Permanent Fund, established with oil dollars in 1976, and using the earnings to finance state government. Walker's plan also called for increased taxes on various industries and reinstating a personal income tax, which was repealed in 1980 in the midst of a flood of oil dollars.[11]

Walker's State of the State address before the Legislature on Jan. 21, 2016, included a review of the Sustainable Alaska Plan and what he saw as the need for it.[12]

"When I filed for governor, oil was over $100 per barrel. When I stood before you at this time last year, the price of oil had dropped 51 percent in 83 days. It was dipping below $50 per barrel. Today, it's $26 per barrel. Some experts predict there will be no rebound for many years," he said. "That means we cannot continue with business as usual. That means we have to change our course."

He went on to recall that the late Jay Hammond, Alaska's governor from 1974 to 1982, as North Slope oil production was taking root, warned of such a predicament.

"While the time to pay the piper may well be some 20 or 30 years away," he quoted Hammond as saying, "if we continue to build a government funded primarily from one resource alone, what a terrible legacy we would leave our children."

"Alaska did not heed Governor Hammond's sage advice, and that day is now upon us," Walker told the Legislature.

. .

IT WAS THAT SPIRIT of facing challenges together, of creativity and resiliency, that Darwin said he discovered when he first came to Ninilchik. It's what Kaye saw when she left the fast pace of New York for a new life in the north. And, in addition to the attempt to safeguard their property, it's what has guided the Waldsmiths' interactions with encroaching development in their area of the Kenai Peninsula.

"It was 49 years ago this last week that I came here, met the principal, and he proceeded to show me around. . . . The community was very welcoming," said Darwin.

Little excuse was needed for a potluck in those days. Holidays were celebrated with community plays. Darwin's early Ninilchik memories include the local fair growing beyond its original venue in the school basement and taking on the task of establishing its own facilities. If something was needed at the school, parents joined forces and local businesses, small though they were, generously contributed.

Inspired by what he saw, Darwin became involved wherever and whenever he was asked—teaching in the classroom, running the buzzer and being the scorekeeper at high school basketball games, or coaching an unfamiliar sport.

"I started coaching wrestling because the kids needed something to do and it was a sport that was available," said Darwin.

The first year, thanks to one young athlete Darwin described as "clever and wiry"—and undoubtedly, despite his modesty, thanks also to the can-do spirit of a fledgling coach—the Ninilchik team went all the way to state competition and brought home a second-place win.

"The next year, we did the same thing again. It was kind of remarkable, kind of unbelievable," Darwin said. Although he had willingly accepted the coaching role, he was nonetheless thankful when the honor passed to fellow teacher John Lindeman two years later.

In her classrooms Kaye was introduced to the lifestyle Darwin had found in Ninilchik. Here were youngsters whose after-school hours were spent shoveling snow off the roofs of their homes, carrying in firewood and hauling water.

"When these kids got home from school, it was all about survival," Kaye said.

Kaye developed a pen-pal connection with students in a classroom in Long Island, N.Y. Through the exchange of letters, Kaye's students experienced city life second-hand and the New Yorkers learned that what they'd heard about the rugged life of rural Alaskans was no myth.

. .

DURING ALMOST half a century of living in Ninilchik, the Waldsmiths have witnessed numerous changes. A different standard of living means less time spent on day-to-day survival tasks. Ninilchik isn't solely the fishing community it once was. Beaches have been closed to the harvesting of razor clams, a resource that fed generations of villagers. The smell of coal, gathered along the beach and used with wood to warm village homes, is rarely detected. With newcomers arriving and descendants of founding families spreading beyond village boundaries, foods that reflect Ninilchik's Russian roots such as the salmon pie known as *pirok* and Russian Easter bread known as *kulich* aren't as common as in times past.

But Ninilchik's people and that survival instinct are still inspiring to Darwin and Kaye, guiding their decisions in the face of the oil and gas expansion that closes in more each day.

"We're just trying to make it somewhat fair to people who are invested in this country and want to remain here," Darwin said. "I'd like to see development happen to benefit everybody. Not just the oil and gas industry."

Apache's roots go back to Minnesota

in 1954, when founders Truman Anderson, Raymond Plank and Charles Arnao, with six employees and the backing of 41 shareholders, supplied $250,000 in start-up capital for Apache Oil Corporation. A year later its first two wells were drilled, both in Oklahoma. The company's interests soon grew beyond oil to include everything from agriculture to dude ranching to shopping plazas, and in 1960 the name was changed to Apache Corporation to reflect its broader scope. Operations were also expanded to other areas of the United States plus Canada, Australia, Egypt and the United Kingdom. In the 1970s, with oil prices rising, Apache's focus was narrowed to energy assets.

In 2010, Apache acquired its first acreage in Alaska. In a February 2013 presentation titled "Riding the Wave of Resource Development," Apache's Alaska general manager, John Hendrix, announced that "with more than 1 million acres, Apache is the largest acreage holder in Cook Inlet." The 2014 objectives for Cook Inlet onshore and offshore holdings, Hendrix said, included obtaining permits for additional seismic activity, completing drilling projects already under way, interpreting seismic data as it was collected and identifying new exploratory well locations.

Apache's 2014 annual report noted company-wide revenue of $13.9 billion, with $12.9 billion in capital expenditures. Alaska rated a one-line reference in the section on North America: "We also have leasehold acreage holdings in the Cook Inlet of Alaska and other areas where we are pursuing exploration opportunities." With recent declines in oil prices and Apache's anticipated reduction in cash flow, the report stated, the company's $2.1–$2.3 billion North American capital budget reflected approximately a 65 percent decline from the previous year.

–McKJ

Debbie

"People have struggled for so long here"

I T TOOK LITTLE TIME FOR WORD TO REACH THE *Homer News* that Apache Corporation had declared a "pause" in its three-dimensional seismic work in Ninilchik in 2012. No sooner had company representatives announced the stand-down to employees and contractors that mid-September morning than the phone rang at my newsroom desk. The call came on the heels of a bad commercial fishing season and a series of storms across the state, including the Kenai Peninsula, so severe that then-Gov. Sean Parnell had declared a statewide disaster.

For those to whom Apache's project meant work during the coming winter, the announcement certainly was anything but good news. The word "pause" offered a spark of hope, however. Perhaps it was merely a temporary delay.

I immediately went down my list of contacts, pulling together a story that explained to residents what was happening. And why.

. .

LISA PARKER, Apache's manager for government relations, was among those I called after learning of the pause. The interruption

was due to Apache's lack of necessary permits, she said. Still
needed was one from the U.S. Army Corps of Engineers that would
allow Apache to place marine nodes on the floor of the inlet to collect seismic data. Apache also lacked a National Marine Fisheries
authorization for the incidental taking of a small number of marine mammals. In addition, she said, because the work Apache was
planning would extend within the Kenai National Wildlife Refuge,
the company still needed to prepare an environmental assessment.
If it was merely a matter of missing paperwork, then what Ninilchik residents wondered that morning was when the paperwork
would be completed. There were cold months ahead and families
to feed.

Lara McGinnis, manager of the Kenai Peninsula State Fair
based in Ninilchik, questioned the status of the fair's contract with
Apache's main contractor, SAExploration. That contract sealed, or
so McGinnis had thought, an agreement for the fair to provide
space for a helipad and an area where the crew could be fed for the
monthly rate of $4,400, paid month by month. The contract was
going to be a welcome boost to the fair's coffers, but beyond that
Lara had hired people based on the anticipated work. Among
those affected by the shutdown was Sharon Dimond, who lost her
project-related employment via the fair. "Everybody was expecting
a little bit out of this whole deal and now it sounds like nothing at
all," said her husband, commercial fisherman Gary Dimond, when
I reached him by phone.

In spite of the economic uncertainty that spread through the
community, Lisa Parker steadfastly maintained that Apache wasn't
deserting Ninilchik. Instead, she told me, the company was preparing to mail cards to property owners declaring, "We've taken a
pause. We'll be back. When, we don't know, but we will be back."

DEBBIE CARY was another of those I called to assess the local impact of Apache's change of plans. For more than 20 years, Debbie has been one of the owners of Inlet View Lodge, a restaurant and bar in the center of Ninilchik. Since 2007, she has been the sole owner.

The Inlet View was originally built from logs in the early 1950s by Wayne and June Bishop. The couple obtained a beer and wine license a few years later and then expanded, adding cabins, laundry facilities, a shower house and bathrooms. The lodge offered food and housing for crews working on construction of the Sterling Highway and for drilling crews during early oil exploration days in the Ninilchik area.

Wayne Bishop died in a plane crash in 1972. During the 1980s, new, more modern buildings replaced the original structure. In 1988, June Bishop died of cancer and the Bishops' daughter, Carolyn, took over operation of the business. After several attempts to sell, it was finally purchased by Debbie, her ex-husband and several family members.

Like Ninilchik's other small businesses, the Inlet View flourishes during the summer when visitors flock to the area. Cars fill the parking lot. The kitchen is busy. Live music, often featuring local musicians, keeps people dancing in the bar on weekends. During the quiet winter months, however, the lodge struggles to stay in operation.

For Apache's project, the Inlet View had reportedly contracted to provide food for the crew. Hearing from others that equipment and foodstuffs had been purchased and employees hired to fulfill terms of the contract, I was eager to know how Debbie was reacting to the uncertainty.

After reviewing her contract, which included her agreement not to discuss its details, Debbie chose not to be quoted. With the possibility that the project would resume, she didn't want to jeopardize relations with a potential customer.

In September 2012, the *Homer News* printed the story about the pause, giving Lisa Parker's explanation as well as reactions from locals affected by the delay.[1]

Five months later, Alaska's then-Sen. Mark Begich announced that Apache had received one of the necessary permits, making it possible to pick up where they had left off. Almost.

"Today the National Oceanic and Atmospheric Administration (NOAA) issued an Incidental Harassment Authorization to Apache. With the IHA, Apache can begin work this spring," Begich said in his February 2013 press release.

For the story on the latest announcement I again contacted Lisa Parker. Did receipt of the permit mean Apache would finally resume its work in Ninilchik?

"I can't answer right now," she told me. "It is our intent to come back. I just don't know when."[2]

. .

As WORK BEGAN on this book, Debbie Cary was one of the people I wanted to interview. More than a year had passed since the Apache project shutdown. Inlet View Lodge was still open. Whatever setbacks the pause created had apparently been overcome. When I contacted Debbie in January 2014, I told her I wanted her thoughts about the overall impact of oil and gas exploration on the Ninilchik area. Contract specifics were not needed, I assured her.

"It impacts us positively" was Debbie's opening assessment as we settled across a table from each other in the restaurant's quiet dining area. When she added, "They're not always forthcoming with information, but a lot of that is confidential," I wondered whether she was thinking of Apache's "pause."

"I do wish they'd spend a little more time letting communities know what's going on. If you're going to be here, talk to us," she continued.

For her part, Debbie has done her best to stay informed about the various oil and gas activities planned for the area. She attends community information meetings whenever they're held, she said, and she listens with an open mind.

"We always have one person yelling about when their chickens died when they did seismic testing," she said, critical of people who "walk in with the idea these guys are jerks, and not even giving the benefit of a doubt for them to explain what they're doing."

Of the various oil and gas companies that have fixed their sights on this part of the Kenai Peninsula, Debbie ranked Hilcorp as the "most honest of any companies I've worked with and at any meetings I've been to." She based her assessment partly on the combination of introductory meetings when company representatives describe upcoming work and meetings at the close of a project that she said offered descriptions of what was actually done.

"What they say at the start and finish isn't the same, but they let you know what their progress is," she said.

As I had heard from others regarding Debbie's interaction with Apache, she described her involvement with the various companies as providing meals for crews that number "anywhere from 15–16 guys to their full-size seismic crews that were 200," she said.

The lengths of the contracts differ. Some even change midstream. She told me of one contract, for instance, that was supposed to last six months but was shortened to 21 days.

"And that's what I said about Hilcorp being the most honest," she said, stressing the importance of knowing what to expect rather than doing her best to prepare for a contract only to have it come to an abrupt end. "One was three months and then had a small extension, so I did that for about four months, but that was a really small one that only had total 20 guys. Then I had one that they said was six to eight weeks and it was six to eight weeks, 80 to 90 guys, breakfast, lunch, dinner."

In addition to preparing menus and purchasing food for that many additional meals, contract requirements sometimes make it necessary for Debbie to hire additional staff.

"I try to keep it local hire. People that are from Ninilchik or native to Alaska. I don't hire anyone from the Lower 48. I wanted people that are here, that I know," she said.

The influx of revenue from oil and gas operations has "been fantastic," Debbie said. "It's allowed us to do some of the things we need to do. People don't understand. We're a small business ... we don't make a lot of money in the wintertime."

Through her eyes, more oil and gas activity in the area is a good thing.

"I hope it brings us an influx of people and an influx of families, especially in this area. Businesses are hurting because there aren't enough people to support businesses. Schools are hurting because enrollments are down, and without enrollment they get less funds. When there are less funds, they have to cut staff and services," she said. Her two daughters attend Ninilchik School. Debbie is the school's volunteer librarian and helps teach second grade reading.

During the 1990s, when an infestation of spruce bark beetles killed more than 1.2 million acres, about 50 percent, of the Kenai Peninsula's spruce trees, logging caused a welcomed spike in the peninsula economy. Logging activities to remove the dead or dying trees have since waned, however, and now "we have nothing," said Debbie.

Poor fishing seasons in Alaska have also increased Debbie's acceptance of oil and gas activities. According to a report prepared by the McDowell Group for the Alaska Salmon Alliance, 1,144 people were employed in the commercial fishing industry in the Kenai region in 2012, which for the report's purposes included Ninilchik, with total gross earnings of $20.7 million—a drop from the previous year, when 1,220 workers were employed and gross earnings totaled $30.6 million, and less than 2013 with 1,204 employees and total gross earnings of $29.3 million.[3]

In 2012, the U.S. Secretary of Commerce declared a federal fishery disaster for areas of Alaska that included the Upper Cook Inlet salmon fishery. The "disaster" designation reached beyond the commercial fishing industry to include sportfish charter operations and lodges, according to Arni Thompson, executive director of the Alaska Salmon Alliance. Two years later, Congress appropriated $75 million for areas hit with a disastrous season. Of setnet fishermen alone, an estimated 443 on Cook Inlet's east side were eligible for payments.

"Commercial fishermen aren't doing that great, and I've seen a lot of (sportfishing) guide services closed up in the last two or three years," Debbie said.

Debbie compared Ninilchik's relationship with the oil and gas industry to the demands of working in Alaska's Prudhoe Bay oil patch. With operations located on Alaska's northernmost shores along the Arctic Ocean, some 12,000 employees from both in and out of state—a monthly average in 2015 of 12,313—fly back and forth from their homes on rotating work shifts that range from week-on-week-off schedules to those of much lengthier durations.[4]

"When you look at the big players, at how much work they're giving to the smaller guys, the trickle-down effect is massive," said Debbie. The increased activity in the Ninilchik area "allows some of these people to stay home, not go to the Slope to work, and actually raise their families."

Will oil and gas exploration and production be a temporary fix for Ninilchik's financial woes? Is it, like the logging that resulted from the beetle infestation, short-lived? Or does oil and gas exploration and production offer more than contracts of a few weeks' duration and jobs that help a few families through a winter? Those were questions Debbie raised but couldn't answer.

"Are they going to stay for the long term or just do this to get the infrastructure set up and then have a mass exit? It's a huge gamble," she said.

. .

Whether long- or short-term, Debbie is certain that the presence of Apache, SAExploration, Hilcorp and other companies in the oil and gas industry is leaving its mark on the area.

"One of the things they stood up and said is that they try to make as little impact as they can," she said, "but if you live next door to where they have a right to work, it's going to affect you."

"But if we don't do exploration," she mused, "we'll run out of the fuels that we need."

It's important to find alternative sources of energy, she said, but "that's going to take time and money we don't have. That won't be our generation. Hopefully, it'll be our kids. Hopefully. But we need to develop those ideas now."

The struggle to remain where she has chosen to live, to keep the business going, to help others in her community and to provide for her family weighs on Debbie.

"People have struggled for so long here. Some days I wonder how to hang on. Some days I think it's time to close the door, time to go away," she said.

However, "as long as the law says yes, it's OK for them to be here, then we all have to readjust the way we're thinking," she said. "Some of us are going to be inconvenienced, but you always have a choice. You can either live with it or go somewhere else. It's not always that easy, but if it negatively impacts you and your feelings, then maybe it's time."

As Debbie said at the beginning of our afternoon together, she believes oil and gas activities in Ninilchik have a positive impact. She also is well aware of the related uncertainties, the risks. The changes that can occur with little or no warning. Food bought, people hired, all the time knowing circumstances could change.

And if they do? On that January afternoon, the Lodge's restaurant tables were empty except for the one occupied by Debbie and me, and the parking lot was devoid of cars. In spite of the seeming lack of business, however, Debbie's kitchen staff was preparing meals for a small natural gas crew working in the area. The "what if" question would wait for another day.

. .

As of July 2015, almost three years after Apache announced the Ninilchik "pause," when or if the pause would end was still unknown. Permits obtained from individual property owners allowing Apache's access to private land had expired and Apache was working to renew those, according to Lisa Parker, as well as to renew the required state and federal permits.

"We may return at some point in the future," she said.

"May return" was enough of an open-ended comment to keep a flame lit beneath any Ninilchik residents or businesses still hoping jobs or contracts might actually materialize. That same month, landowners in the area, my daughters and I included, received letters from the Alaska Division of Oil and Gas stating that SAExploration, Apache's lead contractor in 2012, would be doing seismic work in a 548-square-mile area of the lower Cook Inlet. Was that another sign Apache was picking up where it left off?

Perhaps not. On Feb. 25, 2016, Apache announced its financial and operational results for the previous year, a net loss of $23.1 billion. "While returns are still adequate at the well level, we believe it is better to wait until fully burdened rates of return improve to higher double-digit percentages, before materially increasing our rig count and developing our acreage," John J. Christmann IV, Apache's chief executive officer and president, said in the statement posted on Apache's website.

The definitive answer came on March 3, 2016, when the *Alaska Dispatch News* ran an update under the headline "Apache oil company to shut down Alaska operations." Linking to the *Homer News* 2012 story about the "pause," ADN described Apache as "another victim of collapsed oil prices."

"We recently reduced our spending plans for 2016 by 60 percent from 2015 levels and are focusing our limited dollars on specific international opportunities and strategic testing in North America. Operations we are suspending as a result of the downturn include our Alaskan activities," the article quoted Apache as having announced on March 2. The announcement was not available on the company's website.[5]

My Roots II · Homecoming

S TORMS WERE NEVER STRANGERS IN MY LIFE. That middle-of-the-night battle to stop a howling wind from destroying the inside of our cabin. Waves of turbulent water crashing into the wheelhouse of a halibut boat on Cook Inlet and soaking me as I slept. All the challenges faced as a single mom. However, it was storms beyond my imagining that descended in the 14 years that followed our departure from Ninilchik.

For starters, I hadn't reckoned on city prices, and my new job in Juneau paid barely enough to house and feed us. The girls and I moved into a tiny one-bedroom apartment with neighbors who shared their love for rock music at head-rattling volumes and at what felt like all hours.

Thanks to friends and family, we had a double mattress that we unrolled on the bedroom floor for Jennifer and Emily, a hide-abed couch that I slept on in the living room, a table and three chairs, all on loan. It wasn't what I had envisioned, but was I not my resourceful mother's daughter? Vegetable crates salvaged from a nearby grocery store became shelves for our clothes, and a soup pot and a frying pan both did triple duty in the kitchen.

The wardrobe of jeans and sweatshirts appropriate for Ninilchik was pitifully inadequate for my office job in the state capitol building. Again, the kindness of others: A skirt, a pair of slacks, a blouse, a sweater, a long coat and a pair of dressy boots saw me through that first season.

Jennifer and Emily's plain but sturdy outfits made the transition from homeschooling to public school classroom, but there were other areas where creativity was needed. Their classmates had store-bought lunchboxes, something beyond our budget. A small remnant of quilted material at the fabric store provided just enough for two small drawstring pouches for their lunches— maybe not topical, but pretty. But there were embarrassments the girls didn't share with me until years later. Jennifer's teacher, for instance, assigned the second-graders the task of drawing the floor plans of their bedrooms, an opportunity to learn measurement and spatial relationships. For Jennifer, however, it was an exercise in humiliation as classmates drew canopy beds in elaborately decorated rooms while she had nothing to "measure" except a mattress she shared with her sister on the floor.

Childcare before and after school had to be added to our tiny budget. Finding transportation to and from work was a continual challenge. I was anxious about Emily through her 10-day hospital stay for pneumonia, and it also added financial stress, as did my own hospital stay the summer after we moved to Juneau.

Despite the "resourceful" side I tried to show the world, the frustration of our circumstances weighed heavy on my shoulders, and there were many sleepless, tear-filled nights. Whether the seeming hopelessness of where my decisions had led us can be blamed for my increased drinking, who knows. Maybe it was the familiarity with bars developed during my childhood when customers at Jackinsky's Ranch, many of them relatives, good-naturedly bought soft drinks for "Punky," the nickname my parents gave me. Maybe it was simply my own lack of control and an immersion in self-pity. It any case, to say alcohol made things worse is an understatement.

During those Juneau years, I also gave marriage another try with a co-worker in my job in Juneau. He was a probation officer with a mean streak that served him well when supervising hardened parolees but not when it came up against my own growing anger at life.

The marriage lasted less than two years, with two separations. It was violent, with broken dishes, plants hurled across rooms, name-calling, bruises and physical violence in which we both participated until he took it to a level that, to this day, is beyond my comprehension.

His involvement with another woman finally led to my decision to leave the relationship. I'd had to change my employment as a result of marrying a co-worker, but a job offer with the Alaska Legislature came with an increase in pay. That meant I was able to find an apartment much nicer than the one Jennifer, Emily and I had lived in when we first moved to Juneau.

When I announced to my husband my plans to move, wisely keeping the location to myself, he not only refused to help but also refused to let anyone else in the house to help. The next morning I tearfully explained my predicament to Janelle, a co-worker. By noon, she and all the other women I worked with had taken the day off and were hurriedly stowing everything the girls and I owned in garbage bags and boxes. A moving company pulled up shortly after noon and the movers helped carry our belongings, including pieces of furniture I had acquired since those first months in Juneau, from the house to the truck.

Janelle, a friend indeed, took it upon herself to vacuum as we carried the last items out the door. In a gesture of farewell, she left the vacuum standing in the middle of the floor and one roll of toilet tissue on the dining room table.

By the time the girls were out of school, the move was complete.

A third marriage came almost 10 years later and was even shorter. There was no violence, but neither was there any bonding between him and his two daughters and me and mine. We never lived together, and less than a year after it began it came to an end.

· ·

THE YEARS IN Juneau gave opportunities to expand my sewing beyond drawstring lunch bags and clothes for the girls and me. Jennifer and Emily's dance classes provided excellent training: hours of piecing together dresses of netting, chiffon, velvet and ribbons. Halloween meant designing whatever costumes the girls came up with, from crayons with legs to walking ketchup bottles. One co-worker ordered a gown and jacket to wear to the annual governor's ball. Another asked for dresses to wear to work. For a gala performance a local theater company invited me to create costumes and masks for the entire cast.

For three winters I studied stained glass with John Pabor, a local artist, and fell in love with that medium. I was fascinated by its interaction with light and began designing pieces incorporating shells and slices of rocks and exploring the possibilities of three-dimensional shapes.

The more my creative side found expression, the more I felt at peace—and the more I felt drawn to complete the college education I had set aside years before, in support of my first husband, Steve, obtaining his bachelor's degree.

Five years after Jennifer, Emily and I settled in Juneau, we packed up and moved to Tucson, Arizona, where I had enrolled in the University of Arizona's College of Fine Arts, with an emphasis on fiber arts. With a student loan from the state of Alaska and careful planning, the admissions office and I had worked out a schedule that built on credits I'd picked up along the way and allowed me to finish in two years. I was able to find a renter—another single mom with two daughters—for the condo I had purchased after my second marriage ended, and I eagerly anticipated my return to school, new opportunities for my daughters, and time to become acquainted with my mother's birth state.

Via the Alaska ferry system we set out from Juneau for Washington state. Driving south from there, with the girls lodged in their confined spaces in our loaded-to-the-ceiling Honda Civic, my excitement fueled mile after mile. Our last night in a hotel on the road was in northern California. At a dinner stop the next day I was feeling wide awake and eager to get through Los Angeles during the night, when traffic would be at its lightest. If they were up for it, I told the girls, we'd just keep going until I got too tired to drive. Bless their hearts, they settled in for the duration. By sunrise the next morning, the outstretched arms of saguaro lined the horizon. By 10 a.m., the temperature was rapidly climbing to 104, Arizona's August heat rolling out a very warm welcome mat. By noon, we were in Tucson.

We reveled in that southwest climate. There weren't many days the girls didn't wear shorts to school and we didn't end the day with a swim in the pool at our apartment complex. I loved my classes and the challenge each of them offered. Art—drawing, painting, weaving—became the air I breathed. I interned in an art gallery. I created weavings in trees on campus. I wrote about the experience.

My grades were all As and Bs, but it was writing that scored the highest. My permission was sought for one writing assignment to be used as an example in a forthcoming textbook. An instructor suggested I consider changing majors, but I was intent on a degree in fibers and felt I was creating a future with each warp and weft laid down on my loom.

On the second day of my second year at the university, I learned that a required class had been canceled due to lack of funding. Suddenly, my two-year, stick-to-it plan was threatened. A visit to the admissions office showed no other required classes available. Instead I would have to add another semester, in which, with luck, I could pick up what was needed.

Should I continue with a semester I was unable to maximize? Should I sit out a semester and come back when I could make the most of every dollar I was borrowing? Would it be better to work for the next few months and return to school in January? Would the required class ever come back or would the program requirements change to match what the college actually offered? Clearly, I'd been so intent on the goal that I'd built no wiggle room into the plan.

Increasing my debt without making the necessary progress seemed a poor choice, so I withdrew from school for the semester. When I informed the state of Alaska student loan office, I was told that unless I returned to Alaska in the meantime I would lose my residency status and, as a result, my ability to qualify for the loan program. My budget was already tight. A short-time move to Alaska and back again in January was out of the question. I'd already withdrawn from classes, so the university seemed a closed book.

My enthusiasm to continue with school was burning bright, however, and if I had to return to Alaska I was determined the move would include completing the journey for a bachelor's degree.

Phone calls assured me that my renter would continue to stay in our Juneau condo. Admissions personnel at the University of Alaska Anchorage made it clear that transferring to UAA would mean starting over with a brand new list of requirements. Alaska Pacific University committed to give me credit for the courses I'd already taken and, although they didn't offer a major in fibers, they said they would work with me to build a program around my interests.

By the end of October, the girls and I had moved back to Alaska, only this time to Anchorage, and I had been admitted to APU, with classes beginning in January.

It was not an easy transition, however. A year in Arizona, with its warmth and all that offered, had spoiled us. Bundling up for an Alaska winter did not set well with Jennifer and Emily. Our basement apartment was dark. Making new friends was tough.

While I waited for classes to begin, I filed for unemployment and began searching for a job. Things were tight in Alaska in the mid-1980s, and the only job I could find was part-time work in a fabric store. The pay was embarrassing, even more so when I learned it was less than unemployment.

When the woman renting the Juneau condo called to say she was moving out, the price for my choices felt staggering. As my financial situation—no worse than for many other Alaskans at the time—became obvious, the bank persuaded me to face the truth and go through a nonjudicial foreclosure.

During those months of almost-bare cupboards and an even shorter supply of optimism, I explored my chances of applying for welfare. My ownership of the land in Ninilchik eliminated that possibility, and I was not sorry. I had come to feel as if we were destitute, but that three-acre sliver of land was a reminder that we were not. We still had a place on the planet. I might have to work a little harder to stretch our finances, but our cup was not empty.

The following spring, after our move north, the losing battle to keep the condo, and another semester of classes, the proverbial cupboard was, however, completely swept clean. The girls moved to their father's for the summer and I put our belongings in storage and found work as a bartender in Ninilchik, not the best situation for someone who liked the taste of alcohol as much as I did. It didn't pay much, but the tips were promising.

Shortly after I began serving drinks, however, my brother called to say an alternate was needed for his job in Prudhoe Bay.

He and my sister had both worked in Alaska's oil patch, but it was something I'd never considered. For one thing, I had two daughters to care for. For another, my work experience was definitely devoid of anything that would apply to oil field operations.

Shawn's job at the time was operating the gas shack on the ARCO side of the North Slope oilfield, filling gasoline-fueled vehicles, checking their oil and making sure they had enough window-washer fluid to remove cement-like mud. His schedule was four weeks on, two weeks off. His days began at 4:30 a.m. and ended at 7 p.m. Those long hours meant plenty of overtime. The contractor Shawn worked for needed someone to fill in at the gas shack during his two weeks off and in addition, for the next two weeks, to operate the wash rack, hosing mud off vehicles and equipment. If I wanted the job, it was mine. Boy, did I ever want that job.

A few days later, I was on the ARCO charter, headed north. With Mom's help, I had all the gear I was told to bring: boots, coveralls, clothing to get me through a less-than-warm arctic summer. I'd never even pumped gas in my life, but during the couple of days I had to train with Shawn, I learned what was needed to keep fuel tanks full, vehicles oiled and windows clean.

A two-bunk, second-floor room in a two-story dorm was where I slept. The bathrooms were at one end of a central hallway. The walls between the rooms were so thin that the joke was when the guy next door rolled over, I'd have to tug my blankets back in place.

My alarm went off every morning at 4. I quickly dressed, packed a lunch from the food available in the dining room and was at the gas shack by 4:30, as pickups and crew buses lined up for the start of the morning shift or the end of the night shift. There were four pumps and it was fast-paced. I literally ran from one vehicle to the next, getting the fuel flowing and then doubling back to check oil and pour in the window-washing fluid.

Jobs for women at Prudhoe Bay were mostly indoors—culinary, housekeeping or office work. Outdoors, we were a bit of an oddity and the subject of lots of comments.

One morning, a two-fuel-tank crew bus came in and, after getting the nozzle in place on one side, I ran around to the other side to position the second nozzle. Across the back-of-the-bus emergency exit was taped a colorful, very explicit magazine centerfold. My temper instantly hit "red."

Ripping it off the back of the bus, I reached through the driver's window and shoved it in the face of the operator, screaming at him that he better never, ever come through my gas shack with anything like that again. His denial that he knew what I was talking about only made me angrier. It wasn't until the bus pulled out that I saw the crew behind him laughing at the trick they'd played on both of us.

One ARCO employee made it a habit each morning to ask me, "How'd a gal like you get a job like this?" Recognizing that I was working for a contractor and he represented the boss, I fought back my frustrations and kept a smile on my face until one morning he came through with the same old line, and I couldn't keep my mouth shut.

"On my back, the way every other gal up here got her job," I snapped, hoping to cut him off.

It not only silenced him but me, too. No sooner were the words out of my mouth than I wished I could take them back, wished I could tell him that, like many who were working on the North Slope, it was a family member who had helped me get the job.

Someone else brought his truck for fueling after that and it was months before I saw him again. Our next encounter was in the camp's snack room. I was with my sister, who had come up for a short-term job, when he walked in. The look we exchanged made it clear each of us was embarrassed by our last, brief conversation.

To his credit, he took the initiative to lighten the moment.

"Did you know she got her job up here from her brother?" he said, clearly having found out the source of my employment but completely unaware of whom he was speaking to.

"I know," Risa said. "I'm her sister."

. .

THE PHYSICAL strength needed at the wash rack was a challenge. Using a power hose, I'd climb up, over and under equipment to remove coatings of mud and rocks. My uniform was a cumbersome but protective orange rain slicker and coveralls, steel-toed boots, a hard hat and safety glasses.

One morning a huge 988 loader appeared outside my plywood shelter. The operator climbed down and, with a strong southern accent common in the oilfield, politely asked if I'd clean it for him. Noting the condition of the loader, I told him it would take awhile and suggested he go get a cup of coffee.

Upon seeing his shiny, clean loader and the mud now coating my gear and any exposed skin, the operator apologized and said he'd like to give me something for my hard work.

"Do you smoke dope?" he asked, offering what he apparently thought would make up for my mud-crawling effort.

Marijuana or drugs of any kind were prohibited on the slope. So was alcohol. But that didn't stop the use. Over my years there, I had a roommate with whom I began each day by trotting down to the snack room and getting large glasses of orange juice and ice that we'd take back to our room and combine with vodka. Another roommate and I used to end our days with cups of hot chocolate into which we'd pour blackberry schnapps.

Fifty-gallon drums with the bottoms removed were used as storage for cases of alcohol flown into Deadhorse, the community closest to Prudhoe Bay, and then driven through the security guard checkpoints onto the field for distribution to employees who had placed orders with whoever was brave enough to take the risk.

There was one stretch when that avenue provided me the weekly treat of a bottle of fine champagne. Until the employee was caught during a random search at the guard shack, that is, and was banned from the slope for violating the no-alcohol rule.

Dinners at Deadhorse also were an opportunity to imbibe. One winter, I'd attended so many parties where Crown Royal was the favorite beverage that I used the purple and gold bags the bottles came in to make a quilt for the person who hosted the gatherings in his Deadhorse hotel room.

Before that summer was over, the gas shack and wash rack jobs eventually led to the welcome change of inside office work. Mom came to the rescue, as she had so many times. When it was time for school to begin and my daughters to live with me, Mom volunteered to leave her home during my work weeks and stay with Jennifer and Emily in Anchorage. As much as I wanted to return to my classes at APU, the threat of being as poor as we had been kept me working. During the summers, when the girls were with their father, I worked as long as I could, once putting in a 14-week hitch as an office manager on a pipeline construction job before I took some time off.

When Jennifer began her senior year in high school, I finally left the slope and found work in Anchorage with Shell Western E&P Inc. In addition to being home, the major benefit of that job was that it paid for me to complete my undergraduate studies. In 1992, two weeks before Emily graduated from high school, I finally earned my degree, not in art, as originally planned, but in human resource education.

The year of 1992 signaled big changes in other ways. As the new year approached, I had become increasingly aware of the stomach pain and headaches that hit when I'd begin drinking. Two glasses of wine became too much. The growing discomfort gradually outweighed the welcome flood of warmth and fuzzy comfort alcohol promised. The night of Dec. 31, 1991, I took my last drink and then poured every drop of alcohol on the premises down the kitchen drain. Jack Daniels, Crown Royal, Nyquil, vanilla extract. It didn't matter what it was, if it contained alcohol it was gone. It would be some time before I acknowledged being an alcoholic, but when that realization came, my brother, who had plummeted to the depths of alcoholism and gone through treatment, was there with an outstretched hand.

February 1992 brought another major accomplishment: having an article published in the Sunday edition of the *Anchorage Daily News*. Written for one of my classes at APU, it focused on a moment during a trip my dad and I made to Russia.[1] The positive response I received through phone calls, letters and in person surprised me.

During a visit to Ninilchik several months after it was published, I was introduced to a man who, upon being told my name said, "Oh, you're the writer."

Believing one published article didn't earn me the title of "writer," I replied yes, I liked to write. Then he closed his eyes and, with tears on his cheeks, recited sections of the piece I'd written.

I was startled by this very human demonstration of the link words can create, how their arrangement on a page can speak to hearts and create common ground no matter how different our life experiences may be. It was a defining moment.

That fall, Jennifer and Emily left Anchorage to attend the University of Oregon. If ever there was an empty nest, I was living in it, caught up in the push and pull of what their departure meant. One minute I cried with feelings of abandonment, the next minute I rejoiced at having only my own meal to prepare.

. .

WITH MY DAUGHTERS now making their own lives, with the sanity of sobriety working its way into my life, with the goal of a bachelor's degree finally reached and with my newfound awareness of the bridge between people that writing can build, I made the next big decision of my life: to return, finally, to Ninilchik, to the security offered by the roots from which I grew and toward a future I was eager to explore.

The Arctic National Wildlife Refuge,

an area of 8.9 million acres, was formed in 1960 under President Dwight D. Eisenhower's administration with the aim of protecting the area's wildlife, wilderness and recreational values. Groundwork for the range began taking shape years before that, however. Among those influential in its formation was Alaska biologist Olaus Murie and his wife, Mardy, who in 1956 led a trip to the Sheenjek River valley. The Sheenjek River flows from the Romanzof Mountains within the Brooks Range and continues 200 miles to the south, merging with the Porcupine River near its juncture with the Yukon River. Accompanying them, among others, were biologist, conservationist and author George Schaller and William O. Douglas, associate justice of the U.S. Supreme Court from 1939 to 1975.

"I've traveled in many parts of the world, in the most remote wilderness, and I don't think people in the United States realize what treasure they have, because there is very little remote wilderness left in the world. It is very hard to find a place that is virtually untouched, so the Refuge is really a treasure not just for the United States but for the world," Schaller said after that expedition. [1]

Douglas set aside one chapter of his book My Wilderness to focus on the trip with the Muries.

"Never, I believe, had God worked more wondrously than in the creation of this beautiful, delicate alcove in the remoteness of the Sheenjek Valley," he wrote. [2]

The 1980 Alaska National Interest Lands Conservation Act redesignated the Arctic National Wildlife Range as part of the 19.64-million-acre Arctic National Wildlife Refuge, in the northeast corner of the state. ANILCA also added four new

→

purposes for the area: to conserve animals and plants in their natural diversity, ensure a place for hunting and gathering activities, protect water quality and quantity, and fulfill international wildlife treaty obligations.[3]

Besides the Porcupine Caribou Herd, on which the Gwich'in people have been dependent for centuries for sustenance and livelihood, the refuge is home to polar and brown bears, Dall sheep, muskox and wolves—in all, 37 land mammals, 42 species of fish, and eight marine mammals, according to the U.S. Fish and Wildlife Service, manager of the refuge. Referring to the coastal plain as "America's Serengeti," the Audubon Society reports that more than 200 species of birds come to this area from all 50 states and six continents.

According to the U.S. Energy Information Administration, in 2014 the United States consumed 6.95 billion barrels of petroleum products, an average of 19.05 million barrels per day.[4] An assessment and economic analysis of federally owned lands within ANWR done by the U.S. Geological Survey in 1998 estimated "technically recoverable" oil within a 1.5-million-acre portion of the coastal plain known as the "1002 area" at between 4.25 and 11.8 billion barrels.[5]

A 2005 economic update of that report concluded that "until a systematic subsurface evaluation is accomplished, uncertainty about the size and nature of the resource will remain significant."[6]

Although it referred to the 1998 report as "thorough," Arctic Power, a nonprofit grassroots organization founded in 1992 in support of oil and gas exploration on the coastal plain, also criticizes what the group calls its weaknesses. Using Prudhoe Bay as an example, Arctic Power says, "Estimated recovery from Prudhoe Bay was initially estimated at about 35 percent, but new technology applied since that time has progressed steadily, and recovery is now expected to exceed 65 percent" of the crude oil estimated to be in place. [7]

Lending support for "responsible energy production" in ANWR is the House Committee on Natural Resources, comprised of 44 members including 26 Republicans, among them Alaska's Don Young, and 18 Democrats. Such an effort would, the committee has asserted, "create thousands of jobs, generate billions in new revenue and help reduce our dependence on foreign sources of oil." [8]

ANWR isn't the only U.S. wildlife refuge to receive the gas and oil industry's attention. One-quarter of all refuges, 155 of 575, have been the site of past or present oil and gas activities, according to a Government Accounting Office 2003 study of oil and gas drilling within the National Wildlife Refuge system. In a 12-month reporting period prior to the report's publication, these activities amounted to 1 percent of domestic production. The report found that no assessments had been conducted of the cumulative environmental effects of oil and gas activity on refuge resources.

In Oct. 30, 2003, testimony before the House Committee on Resources, Subcommittee on Fisheries Conservation, Wildlife, and Ocean, Barry T. Hill, director of U.S. General Accounting Office's division of Natural Resources and Environment, said the GAO's study found environmental effects ranging from "small oil spills and minimal debris to large and chronic spills and large-scale industrial development."

"Over the years, new environmental laws and industry practice and technology have reduced, but not eliminated, some of the most detrimental effects of oil and gas activities. In addition, oil and gas opera-

→

tors have taken steps, in some cases voluntarily, to reverse damage resulting from oil and gas activities, but operators have not consistently taken such steps, and the adequacy of these steps is not known," said Hill. "The Fish and Wildlife Service does not have a complete and accurate record of spills and other damage resulting from refuge-based oil and gas activities, has conducted few studies to quantify the extent of damage, and therefore does not know its full extent or the steps needed to reverse it."

Faced with Alaska's overwhelming budgetary shortfall created by the state's almost total dependence on oil and gas revenues, which have been in sharp decline, Governor Bill Walker continued to lobby for the opening of ANWR. The last question taken by President Obama at the National Governors Association reception at the White House on Feb. 22, 2016, turned out not to be a question but rather another attempt by Walker to plead his case. The president responded by stating that his goal was to "make sure that economic development has taken place in Alaska, that folks are being well served, but that we're also preserving the very thing that makes that place so unique and people care about it so deeply." [9]

Coverage of that exchange was published in the Alaska Dispatch News the next day. One day later, KTUU Anchorage Channel 2 ran a story detailing Walker's use of North Slope oil and gas expert Andy Mack to bolster his fight to open ANWR. Mack's contract with the state was up to $50,000 for a three-month period, the station said. Valuing the financial fix that drilling in ANWR could mean above its nearly 20 million acres of pristine wilderness, the cry of the Gwich'in, or the risk imposed on the Porcupine Caribou Herd, Walker said, "There's a prize out there that's well worth the investment of having a contractor come on and help us." [10]

-McKJ

LARA

Two sons, two world views

WHEN IT COMES TO THE KENAI PENINSULA FAIR, whether it's whopper-size vegetables, carnival rides, bucking broncos or livestock auctions, Lara McGinnis is at the center of the action. Since 2005, she has managed this three-day annual event that draws thousands to the Ninilchik fairgrounds.

The fair got its start in 1951, more than 40 years before Lara came on the scene and eight years before Alaska became a state. The late Mary Hawkins and a band of enthusiastic Ninilchik residents, in partnership with the Parent Teacher Association, held the community's first fair in the basement of Ninilchik School. In that initial group was my mom, Alice. I have clear memories of bouncing along in the car as she drove the Sterling Highway's 200-mile stretch of gravel between Ninilchik and Anchorage to gather donations to help make the fledgling event a success. Among the treasures she secured were bags filled with already-popped corn. Stored in our garage until fair time, they were a tantalizing treat too hard to ignore. I poked holes in the bags with my fingers so I could sample the airy, although slightly stale, morsels, an activity I have no doubt directly relates to my current love of carnival food.

The popularity of the fair, once tagged "the biggest little fair in Alaska," has grown over the years. It attracts not only peninsula residents but also vendors and visitors from across the state and beyond. From the initial school-basement affair, the Kenai Peninsula Fair now boasts its own acreage and permanent buildings. Crowds cheer for the wildly popular Kenai Peninsula Racing Pigs as the squealing oinkers tear around a small track. People dance to live music performed on either of two stages and line up for caribou sausage on a bun and corn on the cob hot off the grill. They shout encouragement for cowboys and cowgirls in the rodeo arena and bid on 4H-raised livestock. Many show off their homegrown vegetables, home-baked delicacies and handmade quilts.

Other times of the year, the fairgrounds are used for celebrating weddings as well as honoring the passing of beloved locals. Craft fairs offer gift-buying opportunities. Groups rent the halls for community meetings. Musicians gather to practice and perform. Equestrians teach their skills.

The fair boasts a nine-member board of directors. Lara Mc-Ginnis' title has morphed from manager to executive director; she chairs the fundraising committee and works year round for her pay, which includes a two-bedroom apartment above one of the exhibit halls. As evidence of her dedication, the 2014 fair had 6,877 visitors, 1,165 exhibits and 91 vendors. That year's annual report shows expenses of $172,282 and an income of $180,417. The fair gate, at $53,629, is the single biggest source of income. Second are corporate contributions, coming in at more than $43,000. More than a third of the 23 sponsors listed on the fair's website are oil and gas companies or those directly related to the industry. Among those companies are BP, ConocoPhillips and Hilcorp Alaska.

That's what brought me to Lara's office in January 2014: to ask her about the impact Alaska's oil and gas activities, especially those on the Kenai Peninsula and in the Ninilchik area, have on the fair.

"It is what keeps our doors open. On many levels," Lara told me.

Topping her list of examples was the fair's conversion from heating oil to natural gas a year before we met in her office.

"We went from having fuel oil bills that were close to $6,000 some months," Lara said. "I skipped paychecks and would go four, five months with no income because it was either pay the fuel or pay me. These buildings aren't designed to be frozen . . . we keep them at 40 degrees. We keep it cold to try to save as much money as we can."

That changed dramatically once the fair became an ENSTAR Natural Gas customer. The biggest monthly bill since then was for $900.

"That's sustainability for us," Lara said.

A financial benefit on one hand sometimes leads to a negative impact on the other in an area as small as the Kenai Peninsula, where local support can be mutually advantageous. Lara pointed to the fair's relationship with Home Run Oil, a Homer-based business whose deliveries formerly kept the fair's buildings from freezing during the winter.

"They were the kindest, most generous company I've ever worked with," said Lara. "There were months they had to carry us because we didn't have the money to pay the bill, but they loved the fair enough to say, 'Catch up when you can.'"

Another example of the difference oil and gas companies have made came several years ago when, with fewer visitors than anticipated, the fair's future looked particularly bleak.

"We were contemplating selling off property because we didn't know how we were going to make it. Then BP caught wind of it and they stepped in with a pledge to fund the BP Kids Day the following year and gave us that money right then and there, on the spot. They knew we needed it," Lara said. "Certainly there are some corporations that only care about the bottom line, but the ones I see as sponsors—BP, ConocoPhillips and Hilcorp—have been very generous. They reach out to us and say, 'How can we help give back to the community we're taking from?' That allows me to believe in them, that they aren't just about the bottom line."

Lara was one of those I spoke with in 2012 when Apache Corp. called a "pause" to three-dimensional seismic work they were doing in the Ninilchik area. In a phone interview, Lara told me the pause was cutting off a monthly payment of $4,400 that would have gone to the fair for space that was to be used as a helipad and an area where crews could be fed. The tension in her voice that morning made it clear the disappearance of that anticipated revenue would pose a hardship to the fair. When I interviewed her for this book two years later Lara was more understanding of Apache's change of plans.

"They kept telling us all the way through that this was a hope and please don't bank on it," said Lara, referring to what the fair and other local contractors had been told by Apache. "We believed it further than they wanted us to, but, boy, they kept warning us all the way through."

The pause had come just as the fair was preparing to convert to natural gas, "but everything went haywire because I didn't understand how to make it happen and we were stuck on fuel oil for three months longer than we were supposed to be," Lara said.

When Apache and its main contractor, SAExploration, pulled the plug on their work due to lack of required permits, Hilcorp and its contractor, Veritas, "were so kind, so generous, so easy to work with," Lara said. Hilcorp not only rented fair buildings from which to work but also pulled out temporarily and offered help when a scheduled fair fundraiser took place.

"Again, that renews my belief that oil is not evil," Lara told me.

With 56 percent of Alaska's total budget during fiscal year 2012 and 90 percent of its discretionary spending or general fund coming from the state's oil revenue, Lara was aware that financial aid from the state formed another link between the fair and the oil and gas industry.

"Replaced appliances. Replaced boilers. That was all possible through a grant from the state for community development. Ultimately, that comes from gas and oil revenues," she said. "Having lived in other states, I see our huge dependency, and all of us that live here need to really appreciate it. . . . There are so many opportunities other states don't have because we've taken that relationship and used it to our best advantage."

When I asked Lara how she would respond to others I had interviewed whose lives were being negatively affected because of oil and gas exploration and production occurring next to their homes and, in some cases, on their land, she replied, "I would say I feel for the loss of the aesthetic and privacy and the feeling that their property has been encroached on. I'm a rancher's granddaughter and I understand the privilege of those wide-open spaces, but I would have to say if you look deeply at what communities have gained, I think you can feel OK with that loss. Or at least make peace with it."

There's a stubborn streak, she said, that she believes is shared by Alaskans who prefer "wild places and staying dependent on wells versus city water, a septic system versus city-wide sewer. . . . It's inherent in us to be independent. Those of us that choose a rural lifestyle will never get the monetary benefit from natural gas, but we always get the intrinsic benefit in that it changes our society. There are more opportunities for kids at school or for the guy down the street that's down on his luck," she said. "That's the thing about BP or ConocoPhillips. They are all about taking care of our youth, giving them an outlet, an opportunity to learn about what's going on in the world so they can grow up and contribute to our society."

In addition, she continued, combine exploration and production methods with new technology and you have an industry that is "so non-invasive. It's not like they have to go through and absolutely destroy the world. They came through Ninilchik and we didn't even know they'd been here."

· ·

IN SPITE OF her generous assessment of the oil and gas industry, something about our conversation tugged at Lara's emotions. Shortly after we began talking, her eyes uncharacteristically overflowed with tears. In my 10 years as a *Homer News* reporter, in every interview of Lara I'd done while covering the fair and its fundraisers, Lara had been unfailingly upbeat and enthusiastic. As it turned out, the topic at hand—the impact of oil and gas—had a deeper and more personal connection for Lara than its benefit for the fair. It reached into this single mother's home and involved her biological son, Robert, at the time an eighth-grade student at Ninilchik School, and Ronald, her then-20-year-old adopted son, who is Gwich'in by birth.

"For Robert's seventh-grade research project, he did whether or not to open ANWR," Lara said, referring to the longstanding debate on both the state and national level about whether to open the Arctic National Wildlife Refuge to oil and gas development.

Robert's project examined both pros and cons, Lara said. He took into consideration what potential oil and gas reserves could mean for the country. He addressed support by the residents of Kaktovik, a mostly Inupiat community of 308 residents on the northern edge of ANWR, for oil and gas activities and the possible financial benefit to their village. Robert looked at opposition from those seeking to safeguard this untouched wilderness. He also took into consideration arguments of the Gwich'in people, numbering approximately 8,000, who live in both Alaska and Canada and fear that oil and gas activities within the refuge would negatively disrupt the Porcupine Caribou Herd, whose birthing grounds are on ANWR's coastal plain. The herd was numbered at 169,000 by the Alaska Department of Fish and Game in 2011. Oblivious to national boundaries, it migrates between Alaska and Canada's Yukon and Northwest Territories. For thousands of years, the Gwich'in have depended on the herd for food, clothing and tools.

"Robert's conclusion was that not opening ANWR was based on a lot of what-ifs. Ronald read the report and began to cry because he realized that his brother was supporting stripping his heritage. It was pretty intense, but Robert stood by his report and his research," said Lara. "And I stand by it and all the reasons why we should drill, but I also stand behind Ronald's right to say it shouldn't happen."

Under Lara's roof was a family-sized version of the heated debate that has divided Alaskans, the state and the country for years.

As an employee of Alyeska Pipeline Service Co. in the early 1990s, I was fortunate to drive the 414-mile stretch of the Dalton Highway between Fairbanks and Prudhoe Bay several times. Portions of it skim ANWR's western boundary. The first time I crossed over 4,739-foot Atigun Pass in the Brooks Range and dropped down to the valley below, I was struck by the immensity of the country around me, its incredible beauty, its remoteness.

During an assignment to Alyeska's Pump Station Two, some 60 miles south of Prudhoe Bay, my mind's eye would frequently leapfrog over the pump station's collection of buildings, past the idling pickups, the humming equipment and the pipeline transporting crude oil 800 miles from Prudhoe Bay to the terminal in Valdez, and I would wonder at what was just over that ridge, just beyond those bluffs, just past that horizon.

As of July 2015, 7.2 million acres in ANWR are designated "wilderness," according to Brian Glaspell, refuge manager. He summarized the definition of "wilderness" given in the Wilderness Act of 1964 but urged me to read it in its entirety "because the actual act is pretty short, not like any other legislation you might be familiar with, and it reads almost like poetry."

According to the Act, "A wilderness, in contrast with those areas where man and his own works dominate the landscape, is hereby recognized as an area where the earth and its community of life are untrammeled by man, where man himself is a visitor who does not remain. An area of wilderness is further defined to mean in this Act an area of undeveloped Federal land retaining its primeval character and influence, without permanent improvements or human habitation, which is protected and managed so as to preserve its natural conditions and which (1) generally appears to have been affected primarily by the forces of nature, with the imprint of man's work substantially unnoticeable; (2) has outstanding opportunities for solitude or a primitive and unconfined type of recreation; (3) has at least five thousand acres of land or is of sufficient size as to make practicable its preservation and use in an unimpaired condition; and (4) may also contain ecological, geological, or other features of scientific, educational, scenic, or historical value."

In the 1960s, with Richfield, Shell, Chevron, Amoco and Texaco compiling lease blocks on the North Slope, the possibility of exploring for oil within the refuge received attention. BP attempted to extend its survey area inside refuge boundaries, but the U.S. Department of the Interior said no.[11]

"I'm convinced that petroleum development here, on the coastal plain, is incompatible with the Arctic Refuge's purpose," said Ave Thayer, the refuge's first manager, from 1969 to 1982.[12]

For 10 years, from the early 1950s to the early 1960s, Alaska author Jim Rearden led hunting parties into the Brooks Range and inside the refuge. His two-part article about hunting in that part of Alaska was published in the 1950s for *Outdoor Life* under the title "Best Place to Hunt."

At the same time Rearden was taking hunters into northern Alaska, the North Slope oil and gas industry was taking shape, "and they were running across the tundra with those wheeled vehicles," he recalled when I visited him at his Homer residence in July 2015. That, plus Rearden's stint as an area biologist for Cook Inlet commercial fisheries from 1962–1969, as offshore wells were being developed, shaped his opinion of oil and gas activities in the state.

"They had a habit of tossing garbage and other scrap overboard," he said. "At one time the beach in Seldovia Bay looked like snowdrift with white Styrofoam coffee cups."

When fishermen notified Rearden of sightings of oil on the surface of the inlet, he followed up by chartering a plane to fly over the area. If the fishermen's reports proved true, Rearden would contact the oil producer.

"But I didn't get much cooperation. 'Oh yes, we'll clean it up right away' but they didn't. So I wised up. I didn't bother to call the oil company. I called the news media and, wow, what a difference. When headlines appeared ... about oil in Cook Inlet and they'd quote me, boy, it got cleaned up quick," Jim said.

· ·

AN EXAMPLE OF oil and gas activities within an area dedicated to conserving fish and wildlife can be found closer to Lara and her sons than ANWR. About 60 miles north of Ninilchik, near the community of Sterling, is the Kenai National Wildlife Refuge, roughly 2 million acres in size. Originally named the Kenai Moose Range, it was renamed in 1980 under the Alaska National Interest Lands Conservation Act (ANILCA).

"Executive Order 8979 establishing the Kenai Moose Range on Dec. 16, 1941, could have, but did not close the refuge to oil and gas leasing under the Mineral Leasing Act of 1920," Lynnda Kahn, refuge fish and wildlife biologist, told me in the summer of 2015 in response to my emailed questions. "In view of the economic situation of Alaska at the time, the national defense aspect, and the fact that rich oil deposits were already suspected, the Bureau of Sport Fisheries and Wildlife decided to permit such development if proper safeguards could be found for preservation of the main wildlife features of the range."

On Nov. 29, 1954, Richfield Oil Corp. applied for its first lease of 50,000 acres within the refuge. Two months later, an Anchorage group that became known as the "Spit and Argue Club" filed on an additional 100,000 acres the group would give to Richfield if and when the federal government signed the leases, with the group receiving 5 percent of royalty payments on any oil produced on the leases.[13]

"Members of the club had other motives too. Alaska was still a territory in 1954, and there was a growing impatience to gain statehood. The missing ingredient was a steady and reliable source of state income—the then missing link which oil revenues could provide," former Refuge Manager Jim Frates wrote in 1999 in his historical perspective of oil and gas activities in the refuge.

Throughout the 1950s, refuge staff were inundated with requests for permits to enter refuge land, and an eight-point interim policy was developed to deal with the mounting pressure to allow oil and gas activities.

"From all indications, it appeared as though the Refuge could control and counter any adverse effects of exploration by only insisting on strict compliance with regulations/stipulations. This innocence is further seen in the fact that recommended bonding requirements for oil companies doing work in the Refuge need be only in the $1,000 to $5,000 dollar range!" Frates wrote.

In 1956, U.S. Interior Secretary Fred Seaton signed the Richfield leases. On the day before Thanksgiving, geologist Bill Bishop dug his boot heel into the ground near the base of a lone hemlock tree and said, "Drill here." Although the tree was later destroyed, its significance lived on throughout the Cook Inlet Basin as the "mother lode production strata," the Hemlock Zone, according to Frates.

The drilling of Swanson River No. 1 began in April 1957. On July 23 of that year, an *Anchorage Daily News* headline announced "Richfield Hits Oil."[14] An agreement among Richfield, Standard Oil of California (later known as Chevron Corp.), Union Oil Co. (later known as Unocal) and Ohio Oil (later known as Marathon Oil Corp.) resulted in Chevron being selected as the first unit operator for the Swanson River oilfield, located in the refuge. Subsequent operators were ARCO Alaska Inc. (1986–1992), Unocal Corp. (1992–August 2005); Chevron Texaco (August 2005-January 2012) and Hilcorp Alaska (January 2012–present). The Swanson River field includes 65 well pads; according to the Alaska Oil and Gas Conservation Commission's database, 165 wells have been drilled as of July 2015.

Also within the refuge is the Beaver Creek Oil and Gas Field, begun in 1967. Its operator, Marathon Oil, drilled 11 gas wells. Since Hilcorp took over operation of the field in 2013, seven new wells have been drilled. The field also has two producing oil wells.

"In the 58 years since the discovery well, the Beaver Creek and Swanson River fields combined have reported hundreds of spills of various products including crude, formation-produced water, diesel fuel, methanol, glycol, therminol, solvents, antifreeze, xylene, PCBs and an assortment of other chemicals and compounds," biologist Lynnda Kahn said of incidents that have continued to plague the refuge.

It took 12 years for the impact to be discovered of a January 1972 explosion when a compressor plant blast released propane through a faulty valve. In 1984, it was found that oily gravel from the explosion area and stored in a gravel pit contained PCBs, polychlorinated byphenyls. The gravel had been used to control dust on roads within the field.

. .

DEVELOPED IN the 1940s, PCBs were used in manufacturing transformers, capacitors and other heat-transfer devices. Evidence that PCBs were toxic to humans and wildlife led to the banning of the manufacture and importing of PCBs in the U.S. in 1979. They are now considered probable human carcinogens, and the Environmental Protection Agency's maximum contaminant level (MCL) for PCBs in drinking water is 0.5 parts per billion parts of water.[15]

Gravel from the 1972 explosion had PCB concentrations ranging as high as 220,000 parts per million, according to Lynnda Kahn. From the gravel, the PCBs were subject to release into the air and water, putting workers, wildlife and wildlife-consuming humans at risk. Over a seven-year period the contaminated sites were excavated, the soil gathered and 107,000 tons of the impacted soil incinerated. PCBs remain beneath a building still in use. Completing that phase of the cleanup will wait until the facility is demobilized.

In addition to Swanson River and Beaver Creek, other oil and gas activities within the Kenai National Wildlife Refuge include the Wolf Lake Natural Gas Project, which consists of two gas-producing wells and a 5.5-mile joint pipeline project between Marathon Oil and Cook Inlet Region Inc.; Birch Hill, a 50-year inactive well for which Hilcorp has no plans; the NordAq Shadura Development Project, a 4.3-mile gravel road and 3.7-acre drill pad that, if an initial well test proves favorable, will provide for up to five additional natural gas wells and an industrial water well; and Apache Alaska Kenai Spur Extension and Oil/Gas Exploration Project, a joint effort by Apache, the Tyonek Native Corporation, and Cook Inlet Region Inc., for which state and federal permits are pending.

A 2003 study by the Government Accounting Office found that oil and gas and related facilities in the Kenai refuge had eliminated at least 524 acres of habitat, and associated infrastructure had eliminated an additional 424 acres.[16]

The GAO also reported that managers within the National Wildlife Refuge system "lack sufficient guidance, resources, and training to properly manage and oversee oil and gas activities." That echoes Frates' 1999 observations specific to the Swanson River discovery within the Kenai National Wildlife Refuge.

"Like the local communities, the Refuge staff too was ill-prepared to handle the ominous tidal wave of a well 'oiled' politically-charged industry (that) not only spoke a language unfamiliar to wildlife resource managers, but one whose political pipeline connected directly to Capitol Hill as well as the Oval Office," Frates wrote.

· ·

ACCORDING TO the Fish and Wildlife Service, the question of drilling within ANWR is answered in ANILCA: "Production of oil and gas from the Arctic National Wildlife Refuge is prohibited and no leasing or other development leading to production of oil and gas from the range shall be undertaken until authorized by an Act of Congress."[17]

Because Congressional approval is required, an alternative focusing specifically on oil and gas development was not considered in the final Arctic National Wildlife Refuge Comprehensive Conservation Plan and Environmental Impact Statement released by the U.S. Fish and Wildlife Service in January 2015. As outlined in the Record of Decision documenting the service's decision to adopt the Revised Plan/EIS and signed by Geoffrey Haskett, USFWS regional director, Alaska, in April 2015, the preferred of six alternatives includes a recommendation that Congress expand the "wilderness" designation to the Brooks Range, the Porcupine Plateau and the Coastal Plain Wilderness Study areas. The Plan/EIS will direct Refuge management for the next 15 years or until the document is revised.

"Almost all of the rest of the refuge is now formally recommended for 'wilderness' designation," ANWR Refuge Manager Brian Glaspell told me. "Less than 1 percent is not asked for 'wilderness' designation."

The Revised Plan/EIS also recommended that four rivers be included in the National Wild and Scenic Rivers System and that visitor use management of one river within ANWR be revised.

Criticism of the plan quickly flared among Alaska's top elected officials, as quoted in a Jan. 25, 2015, article in the *Fairbanks Daily News-Miner*.

"It's clear this administration does not care about us, and sees us as nothing but a territory," said Sen. Lisa Murkowski, who chairs the Senate Energy and Natural Resources Committee.

"This decision disregards the rule of law and our constitution and specifically ignores many promises made to Alaska in ANILCA," was Sen. Dan Sullivan's reading.

"Simply put, this wholesale land grab, this widespread attack on our people and our way of life, is disgusting," said Rep. Don Young.

Alaska Governor Bill Walker said that as a result of the plan he would "consider accelerating the options available to us to increase oil exploration and production on state-owned lands."[18]

Undeterred by opposition, on April 3, 2015, President Barack Obama submitted the recommendations to Congress.

"This area is one of the most beautiful, undisturbed places in the world. It is a national treasure and should be permanently protected through legislation for future generations," he wrote.

The president's action brought a wave of relief for Ronald's people, the Gwich'ins.

"When President Obama came out with the CCP, I cried," Bernadette Demientieff, office manager for the Gwich'in Steering Committee, told me by phone from her Fairbanks office. In her mind, the request for an enlarged "wilderness" designation that would keep oil and gas development off the coastal plain is a stay of execution for the Porcupine Caribou Herd on which her people depend.

"If they did drilling up there, the herd wouldn't go there (to birth)," Bernadette said. "Even when our people were starving, we wouldn't go to the coastal plain. That's how sacred it is to us."

According to the Alaska Department of Fish and Game, numerous factors, including changing weather conditions, can alter a caribou herd's migratory routes, and such changes can create problems for the Native people dependent on caribou.[19]

"As human activities expand in Alaska," the department's website says, "the great challenge for caribou management is for man to consider the needs of our caribou herds and ensure that they remain a viable, healthy part of our landscape."

Bernadette Demientieff of the Gwich'in Steering Committee agrees. As to encounters between the herd and oil and gas development she says, "Our caribou would be scared into the mountains where the predators are."

The Gwich'in celebrated a victory on July 21, 2015, when U.S. District Judge Sharon Gleason found for Interior Secretary Sally Jewell and the Gwich'in Steering Committee, defendant and intervenor-defendant respectively, in the state of Alaska's suit seeking to order and direct Jewell to review the state's plan for exploration of oil and gas resources within ANWR's coastal plain, a plan that had been submitted to the U.S. Fish and Wildlife Service regional director in July 2013. USFWS had responded that authorization for an exploration program expired in 1987, that the state's plan did not comply with requirements of the Code of Federal Regulations, and that as a result the service would not review the plan. After the state's request for reconsideration was denied by USFWS, the state had taken its case to the court, only to meet a second defeat.

"This is a great decision for the Gwich'in people," Sarah James, chair of the Gwich'in Steering Committee, said in a press release issued jointly by eight concerned parties: the Alaska Wilderness League, the Center for Biological Diversity, Defenders of Wildlife, the Northern Alaska Environmental Center, Resisting Environmental Destruction on Indigenous Lands, the Sierra Club, the Wilderness Society, and the Gwich'in Steering Committee.

"The Gwich'in are caribou people; caribou are our food, our song, our dance, our clothing, our culture. This is a human rights issue," the statement continues. "The birthplace of the caribou must be protected for future generations. Oil exploration and development would hurt the caribou and threaten the Gwich'in way of life. Today's decision helps protect that. But we need to permanently protect this place we call *Iizhik Gwat'san Gwandaii Goodlit*, or 'the Sacred Place Where Life Begins.'"[20]

Sarah has headed the international Gwich'in Steering Committee—four members from Canada and four from the United States—since 1988 and has made many trips to Washington, D.C., she told me when I reached her by phone at her Arctic Village home.

"That's the resolution we got from our nation, to protect the coastal plain of ANWR," she said. In 2012 the Gwich'in people unanimously passed "A Resolution to Protect the Birthplace and Nursery Grounds of the Porcupine Caribou Herd." The resolution references Article 1 of the International Covenant of Civil and Political Rights, ratified by the U.S. Senate, which reads in part ". . . In no case may a People be deprived of their own means of subsistence." It resolves "that the U.S. President and Congress recognize the rights of the Gwich'in People to continue to live our way of life by prohibiting development in the calving and post-calving grounds of the Porcupine Caribou Herd" and "that the 1002 area of the Arctic National Wildlife Refuge be made Wilderness to protect the sacred birthplace of the caribou."[21]

Safeguarding the territory of the Porcupine Caribou Herd is a charge of nationwide proportion, according to Sarah. It requires educating voters across the United States of the importance caribou have for the Gwitch'in so that Congress will make the right decision.

"Since we can remember," she said, "we always protect the caribou and depend on them for food and everything else. Meat, hides for clothing and boots, arts and crafts. To us, it's human rights. To be who we are."

Now that President Obama has submitted his administration's recommendations regarding the Arctic National Wildlife Refuge to Congress, a waiting game ensues.

"It's a big step, whatever Congress chooses to do. It's still a statement of values and intention and there's meaning there. Even if it doesn't change the rules on the ground," said ANWR manager Glaspell.

Bernadette Demientieff had a message to Lara's son Ronald.

"It's good that he's so young and passionate. Tell him we're proud of him for standing up for his heritage," she said.

· ·

IN SPITE OF THE threat to their culture that Lara's adopted son's people believe they face from the oil and gas industry, Lara maintains she looks at "the ripple effect of what oil brings to our community. . . . It makes the community a better place, contributes to the wellbeing of everyone on the peninsula. . . . Look at the fact that the world needs oil and gas. If we as Americans take it from our resource, we're doing better for our world than if we're taking it from someplace overseas that doesn't have our standards and doesn't care how they leave the world when they're done."

"It's that that makes me believe we can find a way to make the hearts of my sons meld for what's best for our world," she said.

MIKE AND JOANN

"We say 'Thanks, Dad' all the time"

MIKE STEIK IS A TRADITION-BEARER for the tiny dot on the Alaska map that is Ninilchik. In his eighties, Mike is one of the oldest living descendants of the village's original families.

Through his mother, Julia, Mike handsomely inherited the dark skin and black hair of his Alaska Native heritage. From her, he also received the beyond-all-measure work ethic that allowed Ninilchik's founders to survive Alaska's challenging conditions.

Julia's maternal great-great-grandfather Grigorii Kvasnikoff was from Kaluga, a community near Moscow. In Alaska he married Mavra Rastorguev, whose father was Russian and whose mother was of Native descent. In 1847, Grigorii and Mavra and their eight children left their home in Kodiak to settle in Ninilchik.

Twenty-some years earlier, the Russian-American Company had found itself with a problem: what to do with employees no longer capable of working because of age or ill health and weighed down by tax obligations that followed them from the homeland to Alaska. Some of these men had married Alaskan women and established families they did not want to leave. The eventual solution, approved by the Russian government's Minister of Finance, included granting these employees "colonist" status, developing small pensioner communities for them, providing the colonists with houses and basic supplies, and allowing the company to pay any taxes owed by the colonists on their behalf and then collect from them what had been paid.

Beyond that these new "colonists" were expected to take care of themselves and with the help of a small pension to purchase what they could not produce on their own.

They also could supplement their income with the sale of vegetables they grew, fish they caught and the furs from animals they hunted and trapped. Ninilchik was one of those settlements, tucked into an isolated thumbprint of land on the eastern shore of Cook Inlet.[1] Some 40 miles to the south, near the inlet's entrance from the North Pacific, the Russians had established Fort Alexander at the Native village of Nanwalek in the late 1700s. About the same distance to the north was Kenai, also developed by the Russians in that period.

The site chosen for Ninilchik offered a river with fresh water. It was bordered snugly by hills to the north and south that offered some protection from harsh winter winds.

"A land made for man" was how Mikhail D. Teben'kov, a chief manager with the Russian-American Company, described it.

The Kvasnikoff family was dispatched there from Kodiak in the summer of 1847, along with Iakov Knagin, a Lutheran Finn, and his wife and four children.

The village was slow to grow—fifteen years after the Kvasnikoffs and Knagins arrived, the population was between 30 and 35 —but they were a hardworking people, determined to survive. By 1908, it had grown to a community of about 100, according to an article in the *Seward Daily Gateway* newspaper in November of that year. The article mentioned that villagers raised geese, chickens, and vegetable gardens. It mostly focused, however, on the resident herd of cattle. The herd's ancestors, measuring three to four feet in height and weighing between 300 and 500 pounds each, had been imported from Russia 75 years earlier. They were "a very hardy little animal" that preferred living outdoors, gave exceptionally rich milk and were gentle and kind, said the Gateway.

Meanwhile, halfway around the world in the opposite direction from Russia, Mike's paternal grandparents Fritz and Henrietta Steik emigrated with their children from Germany to Wisconsin, then to Victoria, B.C., and finally to Port Angeles, Wash. As an adult, their son Chris struck out for Alaska and eventually made his way to Ninilchik.[2]

In 1924, Grigorii and Mavra's great-great-granddaughter Julia Crawford and Ninilchik newcomer Chris Steik were married. Their union produced 17 children, born between 1924 and 1951. Mike, born in 1934, was the eighth.

Chris and Julia's home, built next to the Ninilchik River, was crowded—one bedroom for the girls, one for the boys. To make the most of the space, Chris built sets of bunk beds that were four high.

Keeping village houses warm during the long winters was a struggle. Temperatures plunging well below zero posed a constant threat. Icy winds barreling across the surface of Cook Inlet and the cold breath of the Caribou Hills to the east managed to squeeze past chinking in the village's handhewn log houses. Villagers did their best to ward off the chill, heating their homes with chunks of coal harvested along the beach in the fall. During the coldest months, one of the village men had the task of going from house to house in the dark of night to keep stove fires burning and villagers from freezing in their beds.

Clothing to protect against the cold required ingenuity. Mike and other villagers of his generation recall "moose socks," a type of footwear created by stripping the skin off the hind legs of a moose carcass. Straw packed inside boots also helped provide some insulation. Too often, however, the cold won out.

"Up until about 15 or 20 years ago, I could pull my toenails right off because they had frozen so many times," Mike once said of the life-and-death struggle against the temperatures.

Drawing drinking water from the river was easy enough in the summer. In winter steps had to be chiseled in the deep snow along the banks and through the river's frozen surface so buckets could be lowered to reach the free-flowing water underneath.

Now a questionable practice because of the heavy metals it contains, each spring villagers spread their winter's accumulation of coal ash over gardens to enrich the soil. In the fall, the dark coolness of root cellars kept the harvested bounty—turnips, rutabagas, carrots, potatoes, cabbages—edible through the winter. But even that system balanced precariously on weather's sharp edge. If it got too cold, the stored food froze and villagers' provisions would be wiped out.

Moose, fish and razor clams rounded out food supplies, as did whatever edible plants could be harvested. Pushki, also known as wild celery or cow parsnip, had to be carefully selected and peeled to avoid its skin-irritating properties. The scent of chocolate lilies isn't pleasant, but the bulb has a rice-like consistency to add to a menu. Then as now, fiddlehead ferns were gathered as soon as they began poking their tender, curled heads through the ground in the spring. And there were plenty of berries.

Even during times of abundance, however, getting enough to eat in the Steik family required a special strategy.

"When it came to dinnertime you better get enough the first helping because you sure weren't getting a second one," Mike recalls.

In addition to the isolation and the battle for survival that bound villagers together, the Russian Orthodox Church offered cohesiveness beyond the spiritual realm.

During the Easter holiday, houses got a thorough cleaning. Bedding was aired outside. Floors were scrubbed. The fragrance of baking *kulich*, Russian Easter bread, filled kitchens.

Christmas was celebrated with "starring," singers going from house to house, entertaining with songs about the birth of Jesus while spinning an elaborately constructed star that recalled the one that led the wise men to the manger in Bethlehem. The singers were welcomed into homes and treated to food and warm beverages.

Easter and Christmas also required new clothing, and Mike's mother sewed new dresses for his sisters and new shirts for the boys. Fabric purchased in Seldovia and brought by boat to Ninilchik was turned into new outfits for the family—or it could be the cloth of a flour sack put to a new use, or the still-good material salvaged from outgrown clothing to be worked into new garments.

The effects of the U.S. purchase of Alaska in 1867 were slow to reach Ninilchik but became strikingly evident in the early 1900s with the arrival of English-speaking teachers. Youngsters were forced to give up the Russian spoken in the village in favor of English, and only English, at school. Slips of the tongue were not tolerated and were "corrected" with physical punishment. Slaps on the hands with rulers. Kneeling on rock salt. Isolation from other students. Mike's recalling such treatment when he began school in the 1940s is evidence of the lasting influence of Ninilchik's Russian roots.

As a teenager, Mike fished commercially with his father, using set nets as well as a fish trap. Another of his first jobs was helping maintain the Ninilchik lighthouse, an installation that warned mariners of Cook Inlet's hidden dangers.

When the opportunity to homestead became available, Mike's parents settled their family on a piece of land on a bluff a mile to the south. Its higher elevation raised them above sea-level winds, but, more important, it offered space for a larger garden.

Life in the Ninilchik of Mike's youth was similar to what it had been when his ancestors first arrived: harsh and demanding. It also was rich and rewarding.

"It was a tough life, but I wouldn't change it for nothing. I wouldn't change my life," says Mike, a trace of Russian accent still evident in his speech.

In 1951, Mike met JoAnn Thebaut, the young woman who would become his wife. Raised in Burbank, Calif., JoAnn had a childhood very different from Mike's. It was big-city living, complete with indoor plumbing and paved streets. But after it was discovered that someone was watching the adolescent JoAnn through the windows of the family's home and then had become so bold as to reach through an open window and try to grab her, the family left city life behind, choosing the remoteness of Alaska over the potential dangers of more populated areas.

"Dad said, 'That's enough, we're moving,'" JoAnn recalls. "Friends of neighbors in Burbank had come up here on their honeymoon and came back enthused about Alaska. They thought it was the most fantastic place in the world. My folks decided that's where we were going to move."

In 1951, the Thebaut family homesteaded in the Kenai Peninsula community of Clam Gulch, some 12 miles north of Ninilchik on the new Sterling Highway that was still under construction. Clam Gulch, a settlement of homesteaders and commercial fishermen, was named for the abundant large razor clams dug from nearby sandy beaches during low tides.

JoAnn's father, William Thebaut, found work as a carpenter in the Kenai area, a 20-mile daily commute to the north over the rough gravel road. The nearest school for JoAnn and her siblings was in the opposite direction, in Ninilchik.

Mike and JoAnn's paths initially crossed that year, 1951, at Ninilchik's first community fair. Mike remembers the fair's beginning for one reason: JoAnn.

"I met her there and won the prize," he says with a loving glance at his wife of more than half a century.

A month after their meeting, the spark between the two became evident not just to themselves but also to Mike's mother, Julia. JoAnn was working for Per and Fran Osmar, commercial fishermen who operated a general store in Clam Gulch. When Julia and Mike happened to stop at the store one day, Julia could clearly see the interest her son had in the young blonde California transplant.

"My mom told me, 'That's for you, Mike'" he says, his voice softening with the memory.

JoAnn began spending more and more time with Mike's close-knit family, drawn to the rhythms of Ninilchik village life. Thousands of miles from her California upbringing, JoAnn embraced this new environment and the rules for running a home she received from Julia. The lessons were often strict, but JoAnn was not dissuaded.

"I got my first bawling out from his mom because I hung a towel next to a pair of undershorts," JoAnn says of her lessons in clothesline do's and don'ts. "Socks were first, then shorts, then T-shirts and towels. Everything had to be in order."

The couple married in Ninilchik in 1955 and began their life together in a cabin they built on Thebaut land in Clam Gulch. The next year, Mike was drafted and assigned to the U.S. Army's Fort Richardson, near Anchorage; the couple set up house in a mobile home they purchased. He served until 1959, plus four years in the reserves.

Their next home was in Torrance, Calif., where their only child, William (Bill), was born. Mike found work first as a laborer and then as a carpenter. The commute, anywhere from 100 to 150 miles a day, became too much, and in 1976 the family settled in Ninilchik. They moved onto a 7.25-acre piece of land that was a portion of Mike's parents' homestead. Mike and JoAnn have lived there ever since.

Almost forty years later, the walls and other flat surfaces in Mike and JoAnn's home are covered with reminders of past times. A scale model of a fish trap Mike constructed. A collection of family photographs. Handmade afghans layered over the backs of chairs and couches. Knickknacks from loved ones on windowsills.

Instead of the thin walls and coal-burning stove of earlier times to keep away the cold, Mike and JoAnn's house is insulated and warmed by an oil furnace. Kitchen cabinets are well stocked with food. Electrical lines have replaced kerosene lanterns. Water from a well is brought into the house through plumbing lines. Laundry is done with a washer and dryer, Julia's clothesline lessons a reminiscence from days gone by. In spite of the challenges that come with age, life for Mike and JoAnn is considerably easier than it once was.

Part of that ease is due to JoAnn's father's forward thinking. In a file cabinet, JoAnn has carefully preserved the correspondence from the 1950s between her father, also William, and Standard Oil of California, dealing with the company's interest in the Thebauts' Clam Gulch property. Not the surface of the land where the family settled after leaving California, but the treasure Standard Oil suspected existed beneath the surface.

For her father's signature granting Standard Oil permission to explore for that treasure, Standard Oil offered to pay the family a small stipend. His thoughts of discoveries that might be made and might somehow provide for his family made him the object of ridicule.

"They laughed at him. No one at that time thought there'd ever be anything," JoAnn says of neighbors who couldn't imagine oil would ever be discovered in this isolated place on the planet. "But he was smart."

The monthly payments from Standard Oil originally amounted to about $10. The checks continued after William's death in 1989, and the amount has gradually increased. Through the years, the agreement between the Thebauts and Standard Oil passed through a succession of companies.

"It started with Standard Oil and from Standard went to Unocal and then to Marathon and then to Hilcorp," JoAnn says of the chain of companies certain something of value lay beneath the surface.

When that certainty finally paid off, the amount of the checks increased exponentially.

"It's been a blessing, those extra dollars. It runs anywhere from $400 to $550 a month, just a little extra something coming in," JoAnn says. "I always thank Dad that he had the foresight to think of this."

· ·

THAT "LITTLE EXTRA something" in the bank didn't ease the couple's shock when they saw the transformation in 2013 to Emil and Fran Bartolowits' property less than a mile south of the Thebaut family's Clam Gulch homestead.

As had I, they recalled the 160 acres the Bartolowitses homesteaded in the 1940s picturesquely crowned with a knoll offering an unobstructed view of Cook Inlet and the volcano-dotted, snow-covered Aleutian Range on the inlet's western shore. They had known nothing of the agreement between Emil and Hilcorp after Fran's death that opened the door for the company to demolish the Bartolowits house, scrape away the knoll's crowning stand of birch and spruce trees, level the area and put down a pad of gravel where a drill rig was erected and support equipment moved in.

"I was just shocked," JoAnn said. "We went up there and everything was gone. There was nothing but a gravel pad. It had all disappeared."

Discoveries made from the Bartolowits site could extend beneath the Thebaut homestead, so it is likely Mike and JoAnn will see their monthly income increase even more. In December 2013, Hilcorp notified all the surrounding property owners that they stand to benefit from whatever discovery might be made there.

"We say 'Thanks, Dad' all the time," said JoAnn. "There's so much buried underneath the ground. It's kind of strange to see all these wells sticking up, but that's the only way they can get to the stuff."

The couple doesn't have subsurface rights to the piece of land they've lived on since 1976. Asked how they would respond if, as happened with the Bartolowits homestead, a pool of oil or natural gas was discovered and drew the attention of a producer, JoAnn's initial response is quick.

"Of course we wouldn't have that right here. It would be hard to see a tower sitting in front, blocking our view," she says.

Then she gives the imagined scenario a second thought.

"It would depend on where they would put it, of course," she says, to which Mike adds, "That's progress."

Already in existence is a pipeline that carries natural gas from wells near Ninilchik to Anchorage, nearly 200 miles to the north. When the pipeline was put in, Mike and JoAnn's neighborhood was notified about the availability of natural gas for their homes. However, there was a catch for having access to that source of energy, a source cleaner and cheaper than the coal, wood, propane or heating oil Ninilchik residents use. They would have to pay to have a distribution line constructed to bring the natural gas from the main line to their homes. The estimated cost to run pipe up the hill in Mike and JoAnn's neighborhood was $13,000 per parcel.

"Believe me, if they brought it up here, we would do it, but it would be impossible for us to pay for them to come all the way up here because no one else wanted to sign up for it," Mike says.

Although the monthly check resulting from paperwork signed so many years ago helps make life a little easier for Mike and JoAnn, it isn't enough for them to afford bringing natural gas into their own home. The world's demand for energy is passing them by. Literally. With the same spirit that allowed Mike to weather eight decades of innumerable changes, more than 60 of those years with JoAnn by his side, this is an irony he has come to accept.

"If we had natural gas up here, we would use it, but it won't be in our lifetime, we know," he says.

The night of March 23, 1989, was a busy one

in Valdez, terminus of the 800-mile pipeline that carries crude oil from Alaska's North Slope oil field. Personnel of Alyeska Pipeline Service Co., operator of the pipeline, were celebrating at a safety dinner. Valdez Mayor John Devens also had called a meeting that evening, to address impacts of oil development on his city. At the same time, the 987-foot Exxon Valdez was preparing to sail to Long Beach, Calif., carrying more than 1.2 million barrels—more than 53 million gallons—of Alaska crude.[1]

"No one anticipated any unusual problems as the Exxon Valdez left the Alyeska Pipeline Terminal at 9:12 p.m., Alaska State Time, on March 23, 1989," according to a report by the Alaska Oil Spill Commission. [2]

Why should there be concerns? In 12 years, more than 8,700 tankers carrying North Slope crude oil had already safely made the transit with "no major disasters and few serious incidents." However, the report continues, "That complacency and success were shattered when the Exxon Valdez ran hard aground shortly after midnight on March 24."

The following details of the incident are taken from the AOSC's findings.

The vessel, with Joseph Hazelwood as captain and a crew of 19, had reached the Valdez terminal on March 22. The next morning it began loading oil for the trip south, with an anticipated 10 p.m. departure. Hazelwood spent the day on the vessel's business plus shopping and drinking alcoholic beverages with the vessel's other officers, according to testimony later given to the National Transportation Safety Board. After returning to the vessel, Hazelwood was informed that the departure time had been moved up an hour.

→

At 9:12 p.m., with the help of two tugs, the Exxon Valdez pulled away from its berth at the terminal. Under the direction of marine pilot Ed Murphy and with a single tug escort, the vessel headed for the Valdez Narrows, seven miles from the terminal. Hazelwood left the bridge about half an hour later, despite Exxon company policy requiring two officers on the bridge during this section of the trip.

Shortly before 11 p.m., the Exxon Valdez reported having passed through the narrows and increasing its speed. In advance of his departure from the vessel, Murphy requested that Hazelwood return to the bridge, which the captain did. At 11:25 p.m., Hazelwood notified the Vessel Traffic Center in Valdez that he was increasing to sea speed and would probably move into an inbound traffic lane if there were no conflicting traffic. The center, a U.S. Coast Guard-installed surveillance system for improving the tracking of tankers in Prince William Sound, confirmed a lack of traffic in the inbound lane.[3] Five minutes later, Hazelwood told the traffic center he was turning the vessel toward the east and reducing speed in order to "wind my way through the ice." Instead of entering the in-bound lane, however, the vessel crossed not only the half mile separating the in- and outbound lanes, but moved beyond the inbound lane.

At 11:52 p.m., the vessel's engines were placed on "load program up," a computer program that increases the engine to sea speed full ahead over the course of 43 minutes. Then, after discussing with Third Mate Gregory Cousins where and how to return the vessel to its designated traffic lane, Hazelwood again left the bridge.

After the captain's departure, Cousins was notified by lookout Maureen Jones that navigational lights that should have been on the port (left) side of the vessel were on the starboard (right) side, their position indicating "great peril for a supertanker that was out of its lanes and accelerating through close waters," the AOSC said. Cousins ordered a course change and notified Hazelwood. When the vessel failed to turn swiftly enough, Cousins ordered another course change. Then, realizing the vessel was in serious trouble, he phoned Hazelwood "and at the end of the conversation, felt an initial shock to the vessel. The grounding, described by helmsman Robert Kagan as 'a bumpy ride' and by Cousins as six 'very sharp jolts,' occurred at 12:04 a.m."

Rushing to the bridge, Hazelwood issued a series of orders in an attempt to free the vessel, with the "load program up" condition continuing for another 15 minutes. The chief mate determined that eight of the vessel's 11 cargo tanks and two ballast tanks had been ruptured and the vessel was losing its cargo. Finally, at 12:19 a.m., Hazelwood ordered the vessel's engines to be reduced to idle.

"We're fetched up, ah, hard aground, north of Goose Island, off Bligh Reef and, ah, evidently leaking some oil and we're gonna be here for a while and, ah, if you want, ah, so you're notified," Hazelwood radioed the Valdez traffic center at 12:25 a.m.

With the vessel facing southwest, the Exxon Valdez's middle rested on a pinnacle of Bligh Reef. Calculations aboard the vessel indicated that within three and a quarter hours 5.8 million gallons had flooded out of the tanker. The nightmare had only begun.

"The response capabilities of Alyeska Pipeline Service Co. to deal with the spreading sea of oil would be tested and found to be both unexpectedly slow and woefully inadequate," the AOSC reported. "The worldwide capabilities of Exxon Corp. would mobilize

→

huge quantities of equipment and personnel to respond to the spill, but not in the crucial first few hours and days when containment and cleanup efforts are at a premium. The U.S. Coast Guard would demonstrate its prowess at vessel salvage, protecting crews and lightering operations, but prove utterly incapable of oil spill containment and response. State and federal agencies would show different levels of preparedness and command capability. And the waters of Prince William Sound—and eventually more than 1,000 miles of beach in Southcentral Alaska—would be fouled by 10.8 million gallons of crude oil."

As the oil spread, so did the number of responders, in what marine biologist Riki Ott of Cordova refers to in her account of the disaster, Sound Truth and Corporate Myth$, as "the initial chaos optimistically called 'spill response.'"[4] Page Spencer, who holds a Ph.D. in ecology and was employed by the National Park Service, found herself on a team charged with monitoring the spill's effects on the Kenai Fjords National Park. On April 5, Spencer and her team were cruising down Granite Island on their way to the Chiswell Islands when they encountered their first oil slick.

"After all the sampling and testing and wondering what it will be like, there is no doubt," Spencer wrote in White Silk and Black Tar, A Journal of the Alaska Oil Spill.[5] "The smell is nearly overpowering. The oil lays on the water surface as a blue-grey coverlet. The surface wind ruffling is dampened. At the edges, the thinner oil gleams in iridescent colors: yellows, blues, green, red. Thick brown globs float in the slick, the leading edge of obscene diarrhea from a sick monster. Only a few sea lions remain on the haulout rocks where we saw hundreds just three days ago. I finally spot 50 swimming in the slick just off the rocks. Kittiwakes and gulls swirl overhead. Although the air resounds with gull screams and sea lion grunts, there seems to be a deathly stillness in the place."

Later that day, having witnessed the struggle being waged by seabirds to survive the thick, slick clutches of this spreading black death, the enormity of that "sick monster" descended on Spencer.

"The ecologist in me has known for days the destruction the oil is causing to entire ecosystems," she wrote. "Not just the death of individual animals, plants, plankton and lifestyles; but the total changes in the energy flow through the system. The dynamic flow of energy, nutrients and life is altered, blocked. The functioning chain of life, ebbing and flowing from one organism to another, from one generation to the next no longer functions." [6]

In Cordova, residents "reacted as if there had been a death in the family," according to Art Davidson, former natural resource planning director for the state of Alaska, in his book In the Wake of the Exxon Valdez: The Devastating Impact of the Alaska Oil Spill. Frustrations boiled over during a community meeting with Exxon representatives who were unable to say what impact the spill would have on fishing. Eager to help, Cordova fishermen formed what became known as "the mosquito fleet," a navy of commercial fishing boats committed to do whatever they could to protect the fisheries. They battled relentlessly to safeguard hatcheries, struggling with resources inadequate for the task. Employing seining techniques, they attempted to use booms to corral the spreading substance. They were, as Davidson quotes Rick Steiner, a Cordova marine biologist and half-owner of a

→

fishing boat, "winging it, figuring out what to do as we went along." [7]

The Exxon Valdez Oil Spill Trustee Council reported that about 10,000 workers, 1,000 boats and an estimated 100 airplanes and helicopters known as Exxon's army, navy and air force were involved in the cleanup effort that Exxon said cost the company about $2.1 billion.[8] Some workers washed beaches a rock at a time. Some operated boats. Some cooked and laundered. Some fought to save oiled animals. Others catalogued carcasses as evidence of what had occurred. [9]

In Halibut Cove, Marion Beck, who has a degree in animal science from Cal Poly University and had worked with seals for the National Marine Fisheries Service, contracted with NMFS to help rehab seals during the spill. In a Homer News story marking the spill's 25th anniversary, Marion recalled an encounter during those four months when Exxon officials arrived at the cove to photograph close-up the release of one of her charges. "Repulsed by the arrogant superior presence," Marion made it clear they were to get their images from shore, rather than near the floating pen. When that proved unsatisfactory to the company representatives, Beck put them in a skiff, rowed them to their helicopter and they left before the release. [10]

In its report, AOSC said, "No human lives were lost as a direct result of the disaster, though four deaths were associated with the cleanup effort. Indirectly, however, the human and natural losses were immense—to fisheries, subsistence livelihoods, tourism, wildlife. The most important loss for many who will never visit Prince William Sound was the aesthetic sense that something sacred in the relatively unspoiled land and waters of Alaska had been defiled."

–McKJ

Robert and Kate

"Nobody ever thinks there's going to be a disaster"

THE 10,016-FOOT, SNOW-COVERED PEAK of Iliamna Volcano dominates the view from a small pullout at Mile 148 of the Sterling Highway. The pullout gives motorists a chance to get out from behind the wheel and stretch their legs, but the real gift it offers is the undisturbed view of Iliamna, the central of five volcanoes in a 150-mile stretch of mountains marking the western shore of Cook Inlet.

To the south, Mount Douglas rises to 7,021 feet. Next comes Augustine Volcano, a 4,134-foot cone that is an island unto itself, then Iliamna, 47 miles across Cook Inlet from the parking area. North of Iliamna, Redoubt Volcano stands slightly taller than Iliamna at 10,197 feet. Northernmost is Mount Spurr, the tallest at 11,070 feet and only 78 miles from the city of Anchorage.

Iliamna was last active in 1953. The puffs of steam or smoke that can be seen rising from the mountain on occasion are not enough to worry Alaska Volcano Observatory scientists or halt air traffic in this part of Alaska, but enough to captivate residents as well as visitors. Iliamna and its neighboring peaks stand as a reminder of nature's striking beauty and tremendous force.

149

Most of the fishing grounds for the inlet's commercial fleet lie to the north, but from this vantage point it's not uncommon in the summer to see halibut and salmon sportfishing charter boats carrying clients to favorite fishing areas. In the winter, when icy winds barrel down the inlet from the north, the water's surface is whipped into a frenzy of whitecaps and swells, making experienced fishing captains thankful to have both feet planted on shore.

Toward the center of the inlet, larger vessels can be seen making their way to and from the Port of Anchorage at the head of this long narrow finger of water protruding more than 150 miles north from the Gulf of Alaska. Also visible are tankers carrying various petroleum products from the Port of Nikiski, located at the inlet's mid-point. Their cargo, coming from oil and gas activities on both sides of Cook Inlet and offshore, are reminders of Alaska's close ties to the oil and gas industry.

In 2013, the residents along this stretch of highway were presented with another reminder when a jack-up drilling rig the Endeavour-Spirit of Independence, was towed to Buccaneer Energy's Cosmopolitan oil and gas lease site and parked two miles off shore. With its arrival, supply boats commenced frequent trips to and from the rig. Artificial lights illuminated an area otherwise lit by the sun as it traveled its daily arch across the sky or the moon sparkling on the inlet's surface. Noise from the rig's engines overpowered the rhythmic sound of waves rolling along the gravel beach.

Although permitted by the state of Alaska to be there only during the workable, ice-free summer months, the rig's presence could not be ignored. Not by people stopping at the pullout to photograph the natural beauty of Iliamna, its neighboring peaks and Cook Inlet. Not by those who live in the area.

For Robert and Kate Boyan, whose small two-story home, topped by a gallery showcasing Kate's beading artistry, is directly across the Sterling Highway, there was one saving grace: trees.

151 • *Robert and Kate*

"We got lucky," said Robert. "If you walk across to the pull-out, (the rig) was right in front of you, but as soon as you crossed the road, we were at just the right angle. Looking out the windows, it was behind a stand of spruce trees."

Still, there was the constant, round-the-clock whining of the rig's engines. And there were the lights, visible through the veil of spruce branches.

"It was lit up like a town," said Robert.

"I could put a scope on it and see people walking around. It was huge. I didn't realize how big until I saw boats getting close to it," Kate said.

She had learned about problems the rig encountered after arriving in Kachemak Bay, near the entrance to the inlet, the year before.

"I'd read it was pretty ancient and they were working on it all that time in Homer. If it couldn't get permits, it probably wasn't in such great shape, but they finally got their permits," she said.

Under contract to Kenai Offshore Ventures and Buccaneer Energy, the Endeavour was brought from Singapore to Alaska with the heavy-lift vessel *Kang Sheng Kou* in August 2012. Built at the Clydebank, Scotland, shipyard, the triangle-shaped rig was rated for a water depth of 300 feet and a drilling depth of 25,000 feet. It had undergone repairs and modifications before its departure from Singapore and was scheduled to be in Homer eight days before being put to work.[11]

Eight days stretched to months, however, due to additional work needed on the rig after its long voyage, the state of Alaska's permitting process, and the U.S. Coast Guard's certification requirements. Archer Drilling Co. was contracted to do the work, including corrections to what had been done at the Keppel FELS shipyard in Singapore. Then, in December, Buccaneer fired Archer for non-performance and after learning that subcontractors were not being paid.[12] Four months later, Archer filed a $6.5 million suit against Buccaneer and Buccaneer countersued, claiming Archer's "misrepresentations, misconduct and delays" had cost Buccaneer $30 million.[13]

Finally—in March 2013, after 218 days at Homer's Deep Water Dock—with work on the rig completed, paperwork in order, and Buccaneer having spent a total of $10.75 million on the Endeavour project, including dock fees, the jack-up rig left Homer for the Cosmopolitan site, towed by two Titan tugs.

After completing its work at that location, the rig was towed to its next assignment in upper Cook Inlet. It wintered near Port Graham, a tiny village near the mouth of Cook Inlet, and finally left Alaska in November 2013, with the aid of the *Zhen Hau 15*, a float-on, float-off heavy-lift vessel.[14]

"We were so glad when it left," Kate said.

. .

THE BOYANS' relief wasn't solely due to the rig and its operations. Even if the Endeavour-Spirit of Independence had been silent, even if it had no lights, even if all it did was sit there, its presence was enough to stir their memories of March 1989. Living at the time in the Prince William Sound fishing community of Cordova, neighbor by water or air to Valdez, the couple had witnessed firsthand the deadly effects of the *Exxon Valdez* oil spill.

And lessons from the spill live on.

"Today some people—usually people who did not witness those dreadful days—say to Alaskans, 'Get over it. It's been 10 years'" wrote Alaska author Sherry Simpson in 1999. "But the *Exxon Valdez* oil spill is nothing people can get over that easily— nor should they. Move on, yes. Make changes, certainly. Forget, never."[15]

A local response to Simpson's reminder could be found at Fireweed Academy, a charter school in Homer. In January 2014, fifth- and sixth-grade students created a "discovery lab" on the spill with the help of teacher Kris Owens, Catie Bursch of the Kachemak Bay Research Reserve, the Prince William Sound Regional Citizens' Advisory Council, the Alaska Department of Fish and Game and private residents. The students presented their findings to the public at the Alaska Islands and Ocean Visitor Center. They created a board game focused on the spill's facts and background: They had studied cleanup methods, the spill's impact on wildlife, beaches and coastal lands, how currents spread the oil, the additional safety provided by double-hulled vessels, the effects of crude oil on humans, and long-term effects of the spill. Their displays included a jar of crude oil recently collected from one of the beaches touched by oil from the ruptured tanks of the *Exxon Valdez*.[16]

Some spill reports attempted to give numbers of wildlife killed as a result of the spill. However, in a 25th anniversary story in the *Homer News*, Wally Kvasnikoff of Nanwalek indicated there was no way to know the total loss. Wally, who worked on beach-cleaning efforts, recalled directions given when a boat filled with bird carcasses.

"They were told to take them out as far as they could go and just dump them," he said, his voice breaking with emotion as he recalled his experience. "Even to this day, I carry a heavy burden from that damn oil spill. . . . I'm 58 now and I've never witnessed anything like that. It was horrifying. It just tore our lives up."[17]

For the Boyans, the Endeavour-Spirit of Independence brought back that tragedy.

"It was awful, with all the dead animals out there. That's what I think of when I look at the rig," said Kate, who still harbors memories of the daily radio announcements of birds killed during the spill and has incorporated images from those days into her beadwork. If such a spill were to occur in Cook Inlet, she said, "It would be a complete disaster."

Of her feelings about the oil and gas industry as a whole, she added, "I don't trust them."

Robert was more reserved.

"I don't know if I totally believe that there's going to be a big environmental disaster, especially here in Cook Inlet," he said.

Since the *Exxon Valdez* spill, steps have been taken to minimize the risk of such an event. The 1990 signing into law of the Oil Pollution Act (OPA) expanded the federal government's ability to respond to spills and provide money and resources as necessary. In part, it created the national Oil Spill Liability Trust Fund, established new contingency planning requirements for both government and industry, and increased penalties for noncompliance. It also created two regional citizens advisory councils, one for Prince William Sound and one for Cook Inlet, to encourage long-term partnerships between industry, government and Alaska's shoreline communities.

The Cook Inlet RCAC has a 13-member board of directors, each representing a specific interest or community. Ten ex-officio members represent various federal and state agencies. Through OPA, the council's activities are directed at improving marine transportation and oil facility operations.

Cook Inlet Spill Prevention and Response Inc., a member-owned nonprofit corporation, is a certified Oil Spill Removal Organization and state of Alaska Primary Response Action Contractor. Its mission is to provide member companies with "the most professional, cost-effective contingency plan assistance and spill removal possible." Members include oil producers active in the Cook Inlet region plus the Kenai Peninsula Borough and Municipality of Anchorage, the U.S. Coast Guard, the Alaska Department of Environmental Conservation and Cook Inlet Regional Citizens Advisory Council. Its inventory of response resources is positioned throughout the area and available to be deployed on behalf of CISPRI members. It has developed a manual of spill response tactics and sponsors training for responders and members.

Cook Inletkeeper, modeled after what began as the Hudson River Fishermen's Association in 1966 and grew into the Waterkeeper Alliance dedicated to the protection of rivers, lakes and coastal waterways on six continents, was formed in 1994 by individuals concerned about ecological changes occurring in the inlet.[18]

The following year, when conservation groups negotiated a settlement with Cook Inlet oil and gas producers over more than 4,200 violations of the federal Clean Water Act, rather than pay penalties the producers chose to provide three years of start-up funding for Cook Inletkeeper. Since its formation, the group says, it has been instrumental in reducing the size of offshore oil and gas leases, establishing a citizen-based water quality monitoring program, conducting studies on oil industry pollutants in Cook Inlet Native subsistence foods, deleting beluga whale habitat from state oil and gas lease sales and prohibiting toxic oil wastes from an offshore oil platform.[19]

In 2003, federal, state and local spill response experts—along with representatives from the oil production and transportation industry, natural resource agencies, and residents of communities around Cook Inlet—joined efforts to address the possibility of oil spills. They developed "Cook Inlet Geographic Response Strategies" as an aid to first responders. Approved by the U.S. Coast Guard Marine Safety Office and the Alaska Department of Environmental Conservation, the strategies were created to serve as "orders" to federal and state on-scene coordinators.

The response document identifies by latitude and longitude areas of marine mammals, anadromous fish, eagle nests, sea otters, intertidal spawning, herring spawning, subsistence, cultural resources, sea birds, waterfowl and shore birds, high recreational use, commercial fishing, land management special designations and coastal habitat. Maps are provided to pinpoint the locations.

The Boyans' home falls within a "Central Cook Inlet Response Zone" in the document. Input from a 27-member working group identified their area as having anadromous fish, eagle nests, seasonal concentrations of waterfowl and shore birds, a high recreational use and a marshy coastal habitat.

In spite of post-*Exxon* efforts, the inlet has not been without its threats. During one 17-year period, 1984 to 2001, 28 incidents spilled 540,580 gallons from pipelines, offshore platforms and vessels.[20]

Some spills are oil-industry related, some are sparked by Mother Nature, some are a combination. Icing conditions at the Kenai Pipeline Dock in Nikiski forced the 600-foot double-hull tanker *Seabulk Pride*, carrying nearly 5 million gallons of petroleum product, away from the dock in February 2006. It drifted half a mile and spilled 84 gallons before grounding on the beach.[21]

As a consequence, Tesoro Alaska Company and Seabulk Tankers Inc. paid the state $429,870 and signed an agreement with the state to address civil oil spill claims and alleged violations of winter ice rules for the inlet.[22] Tesoro also agreed to produce a $35,000 video to train Cook Inlet mariners on ice hazards and made upgrades at its Kenai Pipeline dock.

Nature and the oil industry also collided in 1990 and again in 2009, when eruptions of Redoubt Volcano caused floods of muddy water, ice and logs to course down Drift River. The Cook Inlet Pipe Line Co.'s Drift River terminal, temporary storage for oil from ten Cook Inlet platforms, is located at the mouth of the river. The 1990 event raised safety concerns for facility employees and the security of the tanks that can store as much as 1.9 million barrels of oil.[23]

During the volcano's 2009 eruption, the transfer of oil from the terminal to a tanker was called to a halt, workers sought emergency shelter, and terminal operations were closed. Tankers were used to remove some of the oil and water from the terminal in hopes of reducing the risk. Small wonder that concerns were raised when in 2012, Hilcorp Alaska, one of the Cook Inlet producers, began taking steps to bring the terminal back into use. Among those speaking out against the plan was Bob Shavelson, executive director of Cook Inletkeeper, who characterized the location of the Drift River oil storage facility as "one of the stupidest things you could ever do."[24]

. .

I‌F THE *Exxon Valdez* spill plus living on Cook Inlet and having five volcanoes as neighbors have taught Kate anything, it's that tragedies aren't planned, contingency plans don't always work and worst case scenarios can't be imagined.

"Nobody ever thinks there's going to be a disaster. It just happens," said Kate.

Robert's concern about the oil and gas industry's impacts had more to do with overall changes it could bring and the people who bring them.

"For me the big downside isn't that they're after oil and gas. It's just the way it would feel to live here. . . . The aesthetic that takes over is just not where I would choose to live. . . . I'd want to be out of its way if it has to turn into that," he said.

The five-acre spot the Boyans have called home since 1992 came complete with an old cabin situated about five feet outside a 30-foot right of way. When they began considering remodeling the cabin, they wondered if they should move it back more than five feet. However, choosing to listen to advice they were given that the cabin's location was fine, they went about putting their stamp of ownership on it. They enlarged the cabin's ground floor, added a deck and created the second-floor gallery where Kate sells her beaded designs.

In 2011, the Boyans watched as work got under way to both the north and south of them to construct an underground pipeline that would carry natural gas from wells near Anchor Point to EN-STAR Natural Gas customers in Anchorage. As work progressed and the line drew closer from both directions, the couple realized they sat at ground zero.

"This was the spot where they had to meet," said Robert. "But we didn't think it would have any big impact. They'd cut a clean little slice and drop it in."

What happened was anything but a "clean little slice."

"They just started tearing it up," said Robert of a trench dug six feet from the front of their house. It was so deep the Boyans had to stand on the edge to see people in the bottom. It was left open so long a sheet of metal was installed to bridge the opening so the Boyans could continue using their driveway.

Shortly after the work began, the Boyans' front room window directly above the trench cracked. Robert reported it to someone he thought might be a supervisor, but was told not to expect the operator to replace or pay for it.

"There's no way we could fix every complaint we're told we're responsible for," the man told Robert.

When Kate saw what was happening, she posted "private property" signs on her favorite trees, hoping they wouldn't fall victim to the activity taking place "because when they cut down a tree, they just say, 'oh, sorry.'"

In the midst of the trench being dug, markers from a survey paid for by the Boyans got buried. Robert informed someone he thought might be a project foreman of what had happened.

"He got real snotty and said, 'Here's the number. Call this person. I can't deal with you.' He wouldn't even stop what they were doing and look at it with me," said Robert. "So, for me, the big annoyance about all this gas and oil stuff isn't the gas and oil. It's these contractors they get to do this crap that are snotty and arrogant."

At some point the Boyans were given a letter relating to the pipeline project that had a spot for them to sign. They didn't understand what it signified and they were certain they didn't want to sign it.

"The way it was worded was very official sounding, like it was a requirement of you, but it was just another crappy little tactic that still angers me," said Robert, again pointing out that his objection isn't to oil and gas, but rather the people conducting the work.

Kate asked their neighbor, Norm, if he had received a similar letter. Yes, he had. Did he intend to sign it? Norm's response was equally clear.

"He said, 'Hell, no.' I haven't talked to anyone who willingly signed that thing," said Kate, who shared Robert's perspective on the crews working on the pipeline project.

"I saw a lot more hands in pockets than people doing work. I'm sorry, but it's true how many people were standing around compared to one guy doing something. I don't begrudge anybody a job, more power to you if you can get a job, but please don't bring in all these know-it-alls," she said.

Robert described the overall attitude he and Kate observed as "nasty, confrontational game-playing."

"That's the problem. Not utilizing the gas and oil. It's these people that are the problem," he said.

. .

THE PIPELINE wasn't the first activity that made the Boyans wonder about the location of their home. Five years earlier, a fiber optic cable was being strung in trenches along the right of way. That incident didn't cause damage to their home, however. Nor did it spark heated exchanges with project personnel.

"In some places, like here at our house, at Norm's, and I guess where houses are close to the right of way, they drilled under those stretches so as not to disturb people," said Robert.

Before burying the cable near the Boyans' home, the company doing the work asked if Robert and Kate would sell the 10-foot stretch where the cable was going. The Boyans said no.

"I told them on the phone 10 feet is at my dinner table in my kitchen," said Robert.

The amount of work done in the right of way gave Kate a fleeting idea. "We should have been selling donuts," she said, laughing. "It would have been great."

Moneymaking donuts and arrogant attitudes aside, the Boyans question what will become of this stretch of the Kenai Peninsula.

"If they start putting drilling rigs out here and we're looking at them, this whole curve (of the highway) will turn into something else and we'll be having to make a choice," said Robert. "The experience we have now, which is why we like it here, that'll be gone. So we'll either have to leave or completely change our reality."

· ·

HAUNTED BY THE *Exxon Valdez* disaster, Kate, who has won awards for her artistry, has used her beading to speak out. Around the border of "Bad, Bad, Oil Spill," the center of which calls to mind those injured by the Prince William Sound tragedy, Kate has beaded "March 24th, 1989, the tanker *Exxon Valdez* strikes Bligh Reef, spilling more than a million gallons of crude oil, creating the largest oil spill in North American history."

Ten days after the April 20, 2010, explosion of BP's Deepwater Horizon oil rig in the Gulf of Mexico, Kate posted a photo of "Bad, Bad, Oil Spill" on her web page. Beside it, she reflected on her memories of what she experienced in 1989 and left a message for the people of the Gulf: "My heart goes out to these people, animals and plants who are now facing this reality. Let's think about this and decide. No more offshore drilling."

The history of petroleum in the San Joaquin Valley

Valley can be traced back thousands of years. Asphaltum, a thick black liquid form of petroleum collected from natural seeps, was used by Native Americans living in that area in multiple ways including caulking for canoes and as a protective coating for sinews used in arrow construction.[1] Early settlers also found ways to use the substance, including for waterproofing their roofs. Seeps existed in other areas of California as well, including one in Pico Canyon, where the oil from it was used as an illuminant at the San Fernando Mission.

The first oil well drilled in California was in Humboldt County in 1861, with other wells following at such a pace that 65 companies were involved in drilling for oil by 1865.[2] The first wooden oil derrick in Kern County, one of seven counties in the San Joaquin Valley, was constructed in 1878.[3]

Seven miles northeast of Bakersfield, county seat of Kern County, the Kern River oilfield was discovered in May 1899 by Jonathan Elwood and his son, James, on property belonging to Thomas Means. Within three months 134 wells had been started, the Kern River field was producing almost 1,000 barrels of oil a day, and nationwide attention was fixed on the area. Los Angeles had been the most productive oilfield in California at the time of the discovery, producing 1.4 million barrels in 1899, but by 1901 Kern County was producing twice that amount. Four years later, with an annual production of 17 million barrels, it became the top U.S. producer.[4]

Kern County has continued to lead California in oil and natural gas production.[5] An April 2012 report from the U.S. Geological Survey lists three of Kern County's oilfields—the Midway-Sunset Oilfield and the Kern River and South Belridge fields—among the top 10 producing oilfields in the nation.[6]

The Midway-Sunset Oilfield also is infamous for being the site of the largest accidental oil spill on land in history. In March 1910, a blowout of the Lakeview No. 1 wellhead shot oil and sand 200 feet into the air and began an initial unstoppable flow of an estimated 125,000 barrels a day. By the time it was brought under control that October, the total amount spilled was measured at 9.4 million barrels or 395 million gallons. Less than half was sal-

→

vaged. The majority evaporated or disappeared back into the ground.[7] (BP's Deepwater Horizon oil spill in the Gulf of Mexico in April 2010, which released an estimated 3.19 million barrels of oil and killed 11 people, is the largest accidental ocean spill in U.S. history.)[8]

In the century and a half since it began, California's oil and gas industry has continued to grow. In 2013, its natural gas production ranked 13th nationally and crude oil production third. Its total economic contribution accounted for 3.4 percent of the state's GDP, gross domestic product. It contributed $21.2 billion in state and local tax revenues and 2.1 percent of California's total employment.[9]

Also increasing are concerns regarding the industry's impact on California's people and environment. As new techniques are developed, including hydraulic fracturing ("fracking"), public health worries increase. A 2014 report prepared for the Natural Resources Defense Council estimates that 5.4 million Californians live within a mile of one or more of 84,000-plus existing oil and gas wells. Concerns include air pollution, contaminated drinking water and soils, noise and light pollution, public safety and seismic risks. Kern County is listed as home to 63,430 active and new oil and gas wells, with 330,000 residents identified as the "most vulnerable to pollution."[10]

The Kern County community of Lost Hills, some 40 miles northwest of Bakersfield, and Ventura County's Ojai were highlighted in a January 2015 study.[11] The Lost Hills field is the second largest producer of natural gas in California. In 1976 and 1978, two Lost Hills wells blew out, the wellheads collapsing into large craters. In 1998, a well blew and a resulting fire melted nearby drilling equipment, was visible from 40 miles away, burned 40-100 million cubic feet of gas per day, and burned for 14 days.

Largely dependent on oil, and with oil prices dropping worldwide, Kern County supervisors declared a fiscal emergency in January 2015 due to lower property tax revenue from oil properties.[12] A month later, Kern County was back in the news when the Los Angeles Times reported water officials' discovery of producers dumping chemical-laden wastewater into hundreds of unlined pits being operated without proper permits.[13]

−McKJ

MIKE

"Darkened Waters"

MIKE O'MEARA DIDN'T COME TO ALASKA EMPTY-handed when he made the move from California in 1969. He brought with him some strong opinions about the impact of the oil and gas industry on the environment.

"Over and over again, I had seen places I grew up loving become completely destroyed. Eliminated. Not just tainted—they were gone," Mike said.

He was born in Los Angeles in 1942 and lived there until his move north, but as a youngster he often stayed with family and friends in Fresno and traveled regularly through Bakersfield. As a teen and young adult, he broadened his knowledge of the area by taking off on his own, traveling down side roads to see what he could discover.

"I had a chance to see the oil and gas industry up close and personal and learn a little bit about land law and how vulnerable surface holders are, sometimes even when they have subsurface rights," said Mike. "There have been some pretty nasty battles over that issue, and for the average person, anytime you have to go to court is a huge cost and a devastating emotional turmoil, even if you prevail. I came to Alaska knowing what some of the dangers were."

· ·

"A DIRTY BUSINESS" is how Mike sums up the oil and gas industry. "The product is dirty, the way they have to work it is dirty, and no matter how careful people involved are, it's noisy and disrupts the surface of the ground enormously. There's no way around it."

Throughout the 1960s, before Mike's arrival, Alaska's oil and gas industry had steadily been gaining momentum. In 1960 and 1961, the state's petroleum revenues grew from $10 million to $26 million. In 1962, the Beluga River Gas Field was discovered on the Kenai Peninsula. In 1963, Middle Ground Shoal, the first offshore field in Cook Inlet, was discovered, Chevron's refinery began operation at Nikiski, and Cook Inlet natural gas became a major fuel source for power generation in Southcentral Alaska communities. In 1964, the state held its first lease sale for Prudhoe Bay and the first of 16 offshore platforms was installed in Cook Inlet. In 1966, Alaska Oil and Gas Association was established, its mission to "foster the long-term viability of the oil and gas industry in Alaska." The Granite Point, McArthur River and Trading Bay fields began producing in 1967 and the Kenai Peninsula's Beaver Creek gas field was discovered. In 1968, the Prudhoe Bay oilfield, North America's largest, was discovered in northern Alaska. In 1969, the year Mike made Alaska his new home, the state received a record-breaking $900 million from its North Slope lease sale, a liquefied natural gas plant in Nikiski began exporting to Japan, and the Kuparuk and Milne Point oilfields were discovered.[14]

Mike and his wife, Jan, settled first in Anchorage, where Mike taught in public schools. Both he and Jan were eager to distance themselves from urban living. In 1971, they purchased 120 acres of a 160-acre homestead not far from Anchor River's south fork.

Caretakers kept their eye on the place until the couple moved onto the land full-time in 1976.

"The reason I bought this property is that I could see from living in California what was happening to the natural, open land. It was not a pretty sight. It was being devoured," Mike said. "I looked around over all the western states, trying to find a place to go where I might be able to put together a big enough parcel to prevent that. When I found this homestead, such an outstanding natural habitat and unique place, I thought this has to be it."

It wasn't long, however, before oil and gas came knocking at Mike and Jan's door. In fact, they hadn't even moved onto the land full time before they were approached. Aware of an abandoned gas well about a mile from the property line, Mike said, he wasn't that surprised.

"The guy introduced himself, said he worked for such-and-such seismic company and was going to put a 30-foot-wide seismic trail through the property. I said, well, that's interesting, I'd need all his contact information and who he was working for and my attorney would get back to him," Mike recalled. "He started back-pedaling a lot, but I said on a matter like this I never made any commitment or agreement to participate without legal advice."

Mike also emphasized that the caller "did not have permission to come anywhere near my property. I said I realized there was the issue of surface and subsurface rights, but be assured that I would assert my rights to the greatest amount possible, and he assured me he wouldn't proceed without some accommodation."

After considerable back-and-forth between Mike's attorney and the company intending to do seismic work, the attorney drafted a trespass agreement requiring the seismic crew to enter the property on snowshoes rather than using equipment. It forbade the cutting of any brush or trees, and the seismic company was to leave no trace they had been there other than their tracks in the snow.

The seismic company never did sign the agreement. They did cross Mike's land but were apparently careful to follow all the stipulations spelled out by the attorney. During one of his visits to the property from Anchorage, Mike saw the workers approach on heavy equipment and then leave the equipment outside the property line. When Mike returned the next day he saw their tracks in the snow, and the heavy equipment was gone.

"They hadn't touched any brush or anything," he said.

Although seismic surveys to help locate oil and gas reserves have continued in the area, it wasn't until 2012 that Mike was again approached for access to his property. On that occasion it was SAExploration, working with Apache Alaska Corp., wanting access to conduct seismic work. At first the requests came by mail and Mike simply ignored them. Then, as he was getting ready to head into Homer one day, a woman in a parked truck waved him down.

"Are you Mr. O'Meara?" she asked.

"Yeah, you've got him. What can I do for you?"

She explained she was a local representative for a company interested in doing seismic work.

"She wanted to talk me into letting them on the property. She was very cordial and nice, a local person, but I didn't know her. We chatted a little while and I said I wasn't interested, that I had no reason to let them on my property and I wasn't going to do it. She said OK, and that was the last time I heard from those guys. She didn't get high-handed. She just said OK," Mike said.

That response to his refusal may have been "a complete turn-around from the early days," but Mike has no doubt the tactics would change if the situation were different and access to his property seemed essential to a company. He hasn't been approached since then, "but I wouldn't be surprised at any moment to hear from somebody," he said.

In 1973, under Gov. Bill Egan's administration and while Mike was still making trips back and forth between his Anchorage home and the North Fork property, the state of Alaska sold oil and gas leases located near the mouth of Kachemak Bay. As the lease purchasing companies began unfolding their plans in public meetings, members of the public made their displeasure clear. One such meeting, held by Shell Oil Co. on May 18, 1974, drew an "overflow crowd" of opponents from Kenai, Soldotna, Ninilchik, Homer and Seldovia that included representatives of the fishing community, one cannery, the Alaska Department of Fish and Game, conservation societies, biologists, one state senator, housewives and businessmen.[15] Testimony reflected criticism of "continuing efforts to explore and develop oil activities within the highly documented and productive fisheries area, with little if any attention being given towards studying the effects of the relationship between hydrocarbons and the marine and planktonic life."

Arguing that there was little risk to fisheries, one oil-sponsored biologist said it would take in excess of 30,000 barrels of crude to be spilled for more than 24 hours in the bay before any noticeable damage would occur. And even then, "tides would wash the oil away," he said.

Mike lent his voice to the growing cry of opposition.

"We kept trying to get the commercial fishermen on board, but they didn't want to do it. To a large extent they were 'don't rock the boat' guys. But I'll never forget when the (jack-up drilling rig) George Ferris came into Kachemak Bay without warning, and it came through the mouth of the bay at the time when the crab fishery and shrimp fishery were really big, and it tore up all the crab pots out there. Then the fishermen got real motivated. All of a sudden they were on board, ready to join the fight. That's what it took," said Mike.

In December 1974, seven residents of the Kachemak Bay area filed suit in Superior Court asking that the December 1973 lease sale be set aside. The suit claimed the sale and leasing of tracts in Kachemak Bay were unconstitutional due to inadequate public notice. The suit also claimed a lack of public hearings and studies to determine if the sale was in the public's best interest. Defendants included the state of Alaska, Texas International Petroleum Corp., Standard Oil Co. of California, Shell Oil Co., Texaco Inc., and Union Oil Co. of California.[16]

In January 1975, Standard Oil of California held a meeting similar to the one sponsored by Shell. Again, public testimony opposed drilling activity in Kachemak Bay.[17]

In May of that year, Superior Court Judge Thomas Schultz dismissed the fishermen's suit, saying the plaintiffs knew the Kachemak Bay sites were to be included in the sale prior to the sale occurring.[18] Two months later, the Alaska Supreme Court ordered the lower court to hold a new trial to determine if the state had violated public notice laws when it held the sale, but before that could happen the Alaska Legislature voted to condemn the leases.[19]

By July 1977, negotiations for the state to repurchase the leases were finalized.

"The Kachemak Bay repurchase represents a major turning point in Alaska's history," said Gov. Jay Hammond, who had been sworn into office in January 1977. "It represents the end of an era when we sold our oil resources without any real reflection on the impact of the sale, simply for the purpose of gaining immediate revenue to run our government."[20]

. .

BY THE SPRING of 1989, Mike had finished a studio and had plans to develop his interest and talent in art, working primarily in silver and gold.

"The other thing I was doing at the homestead was experimenting with a lot of low-impact kinds of agriculture, raising strawberries to sell and experimenting with harvesting and selling some of the natural plants that grow out here," he said.

Then the *Exxon Valdez* ran aground on Prince William Sound's Bligh Reef on March 24.

"The *Exxon Valdez* completely changed my life," Mike said. "It ended almost everything I did before that time. It set me on a different path."

Shortly after the spill, the board of directors of the Pratt Museum in Homer directed the staff to create an exhibit documenting response to the spill. Martha Madsen, the museum's director of exhibits and education, approached Mike about putting the exhibit together. Not only had Mike closely followed development of the 800-mile trans-Alaska pipeline that carries crude oil from Prudhoe Bay to the Valdez Marine Terminal, where the oil is loaded onto tankers and shipped to market, he also was known for having strong organizational skills and background as an educator, making him an excellent choice for guest curator.

"I thought long and hard, not really thinking I could do it, but as the spill grew and got worse in those first days, I thought what the hell can I do? What part can I play? Am I going out to scrub rocks? Save wildlife? Where can I play a role? So I decided OK, I'd do it," he said.

It was originally planned to be a six- to eight-week project, but "like the spill, it just kept getting bigger and more out of control," Mike recalled.

At first, gathering information proved more difficult than he had anticipated.

"When I got into it at first, I wondered how the hell were we going to do this. You called everybody up and either they were crying or they didn't want to talk or they were under a litigation gag order and couldn't talk," he said. "It was really intense."

As word spread that the exhibit was being created, interest grew. People began asking what the exhibit was going to include and became eager to share their experiences.

"By early June 1989 we were putting the finishing touches on the Homer exhibition, right at the start of the museum's peak visitor season. Even with barriers and signs barring entry, we found ourselves painting and hanging photographs in the midst of growing numbers of spectators. People simply would not be dissuaded from jumping our barricades and deluging us with questions about the spill. By the time we officially opened, word had reached far beyond Homer and people were coming to town specifically to see 'Darkened Waters,'" Mike wrote in "Let the People Speak," a guest blog for Cook Inletkeeper, the community-based nonprofit devoted to protecting Cook Inlet's watershed and the life it sustains. Mike is one of the founders of the organization, formed in 1994, and serves on its board of directors.

The exhibit, "Darkened Waters: Profile of an Oil Spill," was not only well received locally. Overwhelming public interest turned it into what Mike describes as "an 11-year odyssey." A traveling version of the exhibit opened in 1991, visited 10 states and was viewed by more than two million people in spite of venues being difficult to find due to museums' fear of losing donations from oil and gas companies as a result.[21]

"But finally the Smithsonian Museum of Natural History stepped forward and, after a pilot exhibit at the Oakland Museum, Darkened Waters opened on the Mall in Washington, D.C., in 1991," Alan Boraas, professor of anthropology at the Kenai Peninsula College, University of Alaska Anchorage, wrote of the exhibit. Even that support wasn't enough to win over opponents of the exhibit, he wrote, including Alaska Sen. Ted Stevens, "then a friend of big oil, and Bill Allen, the oil spill clean-up contractor and emerging oil lobbyist, was furious."

Criticism of the exhibit also came from oil company executives.

"I think it's misleading. It ignores scientific fact and presents an inaccurate picture of Prince William Sound as it exists today. It concentrates on the first months after the spill," Dick Cureton, spokesperson in Exxon Corp. U.S.A.'s public affairs department in Houston, Texas, is quoted as saying in the Fort Lauderdale, Fla., *Sun-Sentinel*.[22]

The article included a response from Pratt Museum director Betsy Pitzman.

"In dealing with the industry, Exxon had problems from day one. They didn't like the focus that we had. We needed to present what happened and to withhold judgment," she said.

Mike also is quoted in the article: "I hope (visitors to the exhibit) become more aware of the relationship that petroleum has with our lives. . . . We could reduce our use of oil without altering our lives much at all. I hope people see the direct relationship and make their own decisions."

The permanent exhibit continues to educate nearly 30,000 national and international Pratt Museum visitors annually, according to Museum Director Diane Converse. In an out-of-the-way area of the museum, visitors can sit and quietly view the information included—photographs of the *Exxon Valdez* and Joseph Hazelwood, the ship's captain, and of the exhaustive wildlife and beach cleanup effort, with portraits of the people involved and a map of North America highlighting the 1,200 miles of coastline and 3,000 square miles of water impacted by the spill. A vial holding a small amount of crude oil corresponds to an area on the floor indicating how much water that small amount of oil would cover.

Posted details of the spill include U.S. Fish and Wildlife Service's count of 36,000 dead birds and 1,011 dead sea otters, with estimates running "much higher than the actual body count because

THIS YEAR'S BEST HORROR READ — IT'LL SCARE YOU TO DEATH

many animals probably sank to the sea floor or went off into remote areas to die."

And there are quotes from those directly impacted by the spill:

Heidi Shafford, high school sophomore who helped feed and calm oiled birds: "Until you see a dead eagle, you don't realize how massive they are, or how much we have lost."

Nancy Hillstrand, fisheries technician who organized a remote otter rehab site: "It was like a war zone. We were getting 10 to 30 otters a day, all screaming. . . . They were gouging their eyes out. There was so much pain."

Billy Day, a volunteer who helped with the cleanup effort on the Kenai Peninsula beach of Mars Cove: "We touched the earth and it touched us back. . . . It is this feeling of connectedness that we all must plow back into our lives if this planet is to survive."

Mike credits his work on the exhibit as helping him survive the "community disruption and environmental devastation of Alaska's worst technological catastrophe." The experience made it clear to him that cleanup is impossible for a spill this size. The answer lies in prevention.

His original six-week commitment to the museum morphed into a 19-year relationship during which he worked on numerous projects. His plan to spend his days in his studio creating jewelry was set aside. Instead, his artistic talent has been funneled into the political cartoons he creates for the *Homer News* and for which he has received numerous Alaska Press Club awards.

"My old life ended and I got a whole new persona," he said. "I'm not complaining. I think maybe the best and most important work I ever did in my life was the work I did on the 'Darkened Waters' exhibit, so I can't say that I regret it. You never know what life's going to throw at you. It's always full of surprises."

IN 1998, MIKE and Jan filed for an amicable dissolution of their marriage. Mike has remained on the 120-acre parcel they originally purchased. Jan is on a neighboring 40-acre parcel owned jointly with friends. Finding where they live isn't easy. Directions can't be summoned on a cell phone but come directly from Jan or Mike. Take this road to that road, go past a field, continue to a barn and, well, it's all a bit vague and purposely so. Safeguarding the land is as important now as it was when they first moved to the area.

Once you reach Jan's property, a "No Trespassing" sign hangs across a handmade road that disappears into a thick growth of fireweed, alders and stands of evergreens. The afternoon I visited Mike, after he suggested via cell phone that I walk the 1.25-mile distance if I didn't have a reliable four-wheel drive vehicle and then related recent sightings of brown bears meandering down his driveway, I asked if he would instead meet me at the No Trespassing sign and give me a ride. He kindly obliged, which also gave me an opportunity to look at the land around us while he navigated the narrow, rutted roadway.

After a steep descent bordered by tall stands of trees and thick shrubs, the road makes a sharp turn and opens onto a neatly mowed area encompassing Mike's two-story studio, his house, an assortment of small outbuildings and a scattering of trees, bushes and flowers he has planted. From the living room of his small, multi-level house, windows on three sides offer sweeping views of his property and beyond. Somewhere below, the Anchor River makes its way toward Cook Inlet. Beyond that, on days when the clouds aren't so thick, can be seen houses closer to Homer. And beyond that are Kachemak Bay, Cook Inlet and the peaks of the Kenai Mountains.

Bear sightings, including those shortly before my visit, are common. Game trails crisscrossing the property are frequently traveled by moose. Sandhill cranes visit during their migration north in the spring, some staying long enough to raise their long-legged young before the pre-winter flight south.

Mindful of the wildlife he shares this acreage with, I asked Mike what impact oil and gas development would have if it were to occur in the area. It would be "totally devastating," he said. For the wildlife. And for how he views the land.

Although their purchase of the 120 acres didn't include sub-surface rights, Mike and Jan took steps to protect the surface. In 1991, with the help of the Homer-based Kachemak Heritage Land Trust, they obtained a conservation easement to protect the property's natural resources. The land is located next to and partially within the Anchor River/Fritz Creek Critical Habitat Area, which was identified by the state of Alaska in 1985 as important for fish and wildlife, particularly moose, as well as native flora. In addition, the easement notes the area's contributions to the Anchor River watershed, its valuable open space, "agriculturally significant soil," and scenic vistas. Also pointed out are the area's potential for scientific research and educational activities and the importance of its size as a conservation resource.

The 30-plus page document also forbids any use or development of the land that is inconsistent with the purposes of the easement. As spelled out in KHLT's spring 1992 newsletter, the O'Mearas' conservation easement specifically prohibits subdivision, some commercial uses and clear-cut logging.

Organizations like Kachemak Heritage Land Trust exist in almost every state, according to Mandy Bernard, conservation director for KHLT, a member of the national Land Trust Alliance. Generally set up as nonprofits, land trusts are directly involved with the permanent protection of land and its resources for the public benefit. Perhaps the most well-known land trust is The Nature Conservancy.[23]

"What a conservation easement is doing is essentially donating the majority of your development rights to the land trust," said

Mandy. "The deed restricts how much you can develop the property."

In order to be granted a conservation easement, an area must have at least one of the following four qualities: open space, relatively natural wildlife habitat, historical structures, public recreation value.

"The majority of our conservation easements focus on habitat," Mandy said.

The land trust holds the deed of conservation and is responsible for monitoring and enforcing the terms of the easement in perpetuity. For KHLT, the multi-step process to obtain the easement begins with discussions with landowners and ends with final acceptance of the easement. The easement remains binding, whether the property is sold or passed on to heirs. Since being established in 1989, KHLT has helped preserve more than 3,000 acres of Kenai Peninsula land through conservation easements and land acquisition.

A year after the O'Mearas obtained their conservation easement, a neighboring 20-acre piece of property owned by Gerald and Janet Brookman also was granted a conservation easement. The 40-acre piece of property on which Jan O'Meara has lived since the dissolution of Mike's and her marriage was obtained through a state of Alaska agriculture lease sale. The "agriculture" designation includes built-in restrictions "so I don't think we need a conservation easement," Jan said.

For Mike, the biggest threat to conservation is the change in demographics he's observed in the years he's lived on the Kenai Peninsula.

"There are many more people here that view the land as a commodity, many more that want to see the area more urbanized, more that would be comfortable with much more industrialization. Why that is, who they are, where they came from, that's something for another conversation, but it's real obvious to anybody that has lived here," he said.

For the first 25 years that this was Mike's home, he looked out on "beautiful old-growth spruce forest cloaking the whole area as far as I could see."

When a spruce bark beetle infestation hit the Kenai Peninsula in the 1990s, residents discovered that there was money to be had by selling their trees, which gave rise to a logging industry that peaked and died along with the infestation.

"What that did was leave people holding land where they had done the only thing they could think of at the time to enjoy revenue from it," said Mike. Over the years an influx of people has resulted in properties being subdivided, "and I'm watching new structures go up with great frequency," he said. Looking at an undeveloped area to the south, he added, "I'll probably see that whole side with little houses. It's unbelievable change. For someone who moved out here for the wilderness of it, it's devastating. What it does is reinforce the importance of my work to preserve the area that I control. For all the acreage that's destroyed (as wilderness), it just makes what's left that much more precious and important."

As Mike drove me back to my car after our visit we encountered a pair of sandhill cranes and their colt strolling along the drive. The two adults stepped aside, leaving us room to pass, while the long-legged youngster continued to meander down the road on its own. We slowly followed at a distance, enjoying its presence, before it finally changed direction.

Before finding my way back to the highway, I complied with the most important step in the directions Mike had given me earlier and carefully put the No Trespassing sign back into place.

Prayer for my mother's house

May this house be blessed.
May its walls provide
protection.
May its roof bring
inspiration.
May its foundation
conduct the power of the Earth.
May its windows channel
the brilliance of the sun,
the magic of the moon,
and the beauty of the
changing seasons.
May all who stay here
feel at peace.

—Jennifer Long Stinson

My Roots III · Building

THE NINILCHIK I WENT BACK TO IN THE SUMMER OF 1993 was very different from the one I'd left in 1979. The biggest change was loss of the piece of land and the rebuilt cabin the girls and I had called home. Throughout the years, its existence proved a reminder that no matter what challenges arose and no matter where we were, we still had a home. It was small and simple but all ours.

Until my father saw it.

In spite of the changes—new location, new roof, a slightly different floor plan—it was made of the same logs that had been cut and shaped during Dad's childhood to fit perfectly one on top of the other. These same walls had sheltered him, his parents, his siblings. His brother's name was carved into the front, a simple record of a young life tragically ended. The saw cuts evident on the interior walls bore witness to the hard work needed to survive in this demanding environment.

Although he was still working as a captain for the Alaska Marine Highway System and had said more than once that he had no plans to return permanently to Ninilchik, Dad began spending increasing amounts of his vacation time at the cabin, drawn, I'm sure, by its reminders of days gone by. Each of his stays brought "improvements." Pieces of carpeting were tacked over plywood floors I'd spent hours sealing and polishing to a shine. A log I'd salvaged from the beach and displayed in the center of the cabin as support for the loft was boxed in, its beautifully grayed, water-worn surface hidden by milled lumber onto which were tacked electrical switches. Electric baseboard heaters provided a source of warmth much easier to control than the constant attention needed by the wood stove.

With each addition, Dad gradually reclaimed his childhood home. By the time I learned of the changes taking place, the momentum was too powerful to reverse.

The result, with Mom as mediator, was for me to relinquish ownership of the cabin and the land it sat on in exchange for a similarly sized neighboring piece of the homestead.

Thus when I returned to Ninilchik in the summer of 1993, lacking a home of my own, it was once again to Mom's hospitality. Her 14- by-16-foot cabin had expanded over the years to include a tiny basement, which provided me some privacy. My bed and a table, the only pieces of furniture I brought with me, and boxes containing various other possessions filled the space. Disheartening as it was to see the old log cabin and not have access to it, I was back on the homestead. I was home.

The move meant leaving my job with Shell in Anchorage. However, that connection provided temporary employment at the company's Kenai office. The 50-mile-each-way commute was a small price to pay for the move I'd longed to make.

During my free hours, my dog, Chipper, and I gradually got acquainted with our new three-acre parcel of family history. It was covered with birch, spruce, cottonwood and patches of alder, through which filtered sounds of the inlet, a view of the range of mountains on the inlet's far shore, and the blindingly beautiful rays of light that flooded the sky at sunset. Two tiny natural clearings offered space to build a small structure without having to cut down any of the trees. One clearing was larger than the other and it was on that piece of land I focused my attention, spending hours becoming familiar with its contours, its smells and the sounds that flooded the enclosed space.

Having lost my previous claim to a piece of the homestead, returning felt risky. Would I rebuild only to lose what I built? Having found my way back, would I then find forces outside myself redirecting my plans? Was the vision I was creating for my future subject to others' wants and wishes?

Communing with the land was my way of quieting those questions. Allying myself with this piece of the planet offered solid footing. Would it be appropriate for me to live here, I asked the energy I sensed was present. Would *it*, for lack of a name, allow my presence? Could I dig into its surface in order to build a home of my own? Would it offer water for my day-to-day living? Would the moose and bear that crossed it and the birds that made their homes there accommodate my existence or would I be considered an intruder?

During this time the need to leave the trees in place became abundantly clear. After going to bed one evening—I'm not sure if I'd only closed my eyes or had already fallen asleep—I experienced walking through the tall grass toward the larger of the two clearings. A curving line of birches marked my way, leaf-filled branches arching above me. Fragile, papery bark spiraled on their trunks.

Suddenly, a line of beings stepped into my field of vision and began to approach. I don't know if they had been hidden from view because they were standing behind the trees or if they emerged from the trunks, but in any case there they were, their eyes fixed on me as they moved in my direction.

The shock of their presence caused my eyes to fly open and brought me upright in bed. Their appearance is hard to describe.

Perhaps they were light. Perhaps they were filled with light. All I knew then and recall now is the unquestionable awareness that they were of the trees and that the responsibility for their safety was in my hands. Cutting down the trees in order to live on this land was, without a shadow of a doubt, not an option.

· ·

In September, without a house of my own to which I could invite friends and family to help celebrate my return to Ninilchik, I sent invitations to an "earth-warming." Past co-workers and friends from Palmer and Anchorage drove hundreds of miles to attend. My daughter Jennifer was able to get away from work to be on hand. An assortment of other family members joined in. Even my third ex-husband, Sam, and his father were part of the celebration.

With shovel in hand, I officially broke ground, proclaiming the larger of the two little clearings as the place I would make my home. A campfire was lit and, as we soaked up its warmth, each of those present expressed a thought, a quote or a prayer honoring their connection with Mother Earth.

Picnic tables outside Mom's house were covered with food, and when a sprinkling of rain began to threaten the gathering, blue tarps were quickly strung between the trees to keep us dry.

"A slice of heaven," wrote a friend in an album of photos commemorating the day.

· ·

As the months went by and Mom's basement walls began to close in, I felt the increasing need to get into a place of my own. When in February 1994 I was offered work on the federally mandated electrical inspection of the trans-Alaska pipeline, I quickly accepted. Work on a project of this magnitude associated with Alaska's oil industry meant long hours, which meant increased pay and a way to finance construction of a cabin. Mom was as generous as she had been when the girls and I lived with her in the late 1970s, but it isn't always easy—for either generation—for an adult child to live with a parent, so the opportunity to work a rotating schedule also was a welcome change.

In the summer of 1995, almost a year and a half later, I had settled on a plan for the cabin and had enough in a savings account that I felt ready to take the next step.

Deciding on the cabin's appearance was a process similar to seeking the land's permission to build. An idea would emerge, I would sketch it, imagine living in it, ask the clearing for guidance and eventually discard it and start again. Finally a simple 20-by-20-foot structure that felt comfortable to me and appropriate to the space took shape on paper. Each of the four sides had gables, which joined at the center. The main floor included a 20-by-12-foot living room area, a 6-by-8-foot kitchen, an 8-by-8-foot office I envisioned as the perfect writing space, and a 6-by-8-foot bathroom. Above was a 12-by-20-foot sleeping loft. It was tiny but would fit in the clearing perfectly.

Wanting to be certain I was on the right track, I envisioned living in it under all foreseeable circumstances. Winter, summer; morning, evening. Eventually it occurred to me that the design was the same as the St. Nicholas Russian Orthodox Chapel in Kenai. No longer in use, the chapel had been constructed during the early part of the 1900s of handhewn logs in the same style as our homestead house. To verify the similarity, I visited it and was immediately immersed in the feelings I'd imagined I would have inside the cabin I had drawn on paper.

Next, I made a to-scale model of the design with poster board so I had something three-dimensional to carry with me. Again I gave time for it to either fade from my imagination so I could move on to the next idea or for it to continue solidifying as a plan.

When the thrill didn't diminish, I shared my design with Uncle

George, Dad's younger brother. He thought the roof might be a challenge but provided the names of two builders he thought could frame it in. After studying the poster-board model, they said they'd never built a roof like that and wanted time to think it over. Two weeks later, they called to say they were willing to take it on. The price they quoted matched what I'd saved to that point. Construction was going to begin!

The day the dump trucks arrived to put in the driveway so the builder could start excavating for the foundation was momentous. For one thing, it seemed so long in coming. For another, it was the first time I broke my commitment to the trees, an unfaithfulness I've never been able to resolve.

The first load of gravel arrived and, carefully winding his way through the trees, the driver emptied the load and left for another. Upon his return, he decided the opening in the trees was too narrow for him to safely maneuver to the site. His solution: Cut down a spruce tree he now decided was blocking the way. No need, I argued. Just curve around it, dump the gravel and curve back out. It had been done once, it could be done again. No, he said. It was impossible. Unless the tree was removed, he wouldn't deliver the gravel. End of the work. End of the driveway construction and cabin building. All the planning, working, and longing for this day came to a standstill as he waited for my decision.

Surely this couldn't be the end. It was just one tree, I told myself. Besides, it was a spruce, not one of the birch trees I'd seen in my dream.

I gave the OK, the tree was cut and the work proceeded. By the end of that summer, the cabin was framed in and my bank account was empty.

The cabin's shape led to the idea of having a Russian Orthodox-like dome added to the peak of the roof in much the same fashion as the one topping the Kenai chapel. John Pabor, the artist with whom I'd studied stained glass in Juneau, was now teaching sculpture at a university in New Mexico. I sent him pictures I'd taken of dome-topped churches on the trip to Russia with Dad several years earlier, and pictures of churches in Alaska, and asked if building a dome for the cabin was something he thought he could do.

Using copper and brass, John and his students spent months creating and piecing together a dome. As they worked, word spread to my family and friends. Someone suggested the dome be a time capsule, with items given to be placed inside it. Mom put together a history of her side of the family. Jennifer wrote a "Prayer for my mother's house." Emily, who was working on the North Slope at the time, gathered a bouquet of Alaska wild cotton. Other items arrived, each one carefully chosen or created by the contributor.

Finally, the dome was complete. It and John arrived in Ninilchik, and one weekend during the summer of 1996, with the help of my friend Kathleen Bielawski, we raised it to the top of the cabin, tucked all the offerings inside, and bolted it into place.

By the next spring I was working a week-on-week-off schedule as the administrative support for Alyeska Pipeline Service Co., operator of the 800-mile-long line that delivers oil from Prudhoe Bay to Valdez, where it is loaded onto tankers for delivery to out-of-state refineries. My assigned location was Pump Station Two, a small camp 70 miles south of Prudhoe Bay. Wildlife was abundant. Caribou sightings were common, and a herd of musk ox foraged a few miles south of us. A nearby hill offered a perfect spot for watching the aurora on clear winter nights. The remote setting created a family-like closeness among those of us who worked there.

One day, after a co-worker listened to me complain about the time it was taking to save enough to finish the cabin and the frustrations of being a 40-something kid living with my mother, he suggested exploring bank financing. Previous hard times had forged my plan to pay for the cabin as I went. A loan had never been a consideration. However, the pressure to live on my own won out. On my next R&R I contacted a general contractor and a bank and before long work was under way to bring the cabin to completion.

The small team of Batir Construction spent time to understand my vision for life on the homestead. The result was birch for interior trim and for the custom-made cabinets, granite for the tiny kitchen countertop. A spiral staircase minimized space needed to access the sleeping loft. The bathroom sink, glazed with the design of a salmon, came from an Oregon artist. Windows on every wall flooded the cabin with light and kept me connected with the out-of-doors. Lighting was designed to maximize the height of the ceiling and encourage a sense of roominess.

To one side of the cabin, the builders improved on my design with an 8-by-8-foot addition. This would provide a *calidor*—a sort of mud room—at the ground floor entry, with shelves for storage and space for a stackable washer and dryer, and on the second floor, much-needed closet space. I was thrilled to see this extra room in the works until I arrived one evening to check on the day's progress.

"We know how you feel about the trees, but this one will have to go," Jeff told me, pointing to a birch tree. "It's too close to the wall."

I was devastated. My decision to cut down the spruce tree during construction of the driveway still haunted me. This was a birch tree, not a spruce. How could I rationalize its removal? As hard as I tried to come up with an answer, the truth was that cutting it down couldn't be rationalized, but it would have to go if I wanted the addition. And I wanted the addition. If it was to be removed, I decided, it had to be at my hands. I was the one who had promised to save the trees and now I was the one deciding this tree had to go.

I borrowed my cousin's chainsaw and cut it down.

The cabin and the birches, spring 2004. Photo by J.R.B. Pels.

BY THE END OF that summer, the cabin was complete. After years of waiting, I was in my own home. On the homestead.

I still needed an income, however. When Alyeska's move to a more automated pipeline brought the closure of Pump Station Two, I took an internship with the company's training department, based in Fairbanks. My schedule continued to be a week-on-week-off rotation, but the weeks "off" included more and more meetings in Anchorage and less and less time at the cabin.

At last the call of the homestead coupled with the long-resisted dream of focusing on writing persuaded me to leave the oil industry. My 50th birthday was less than a year away, and I was afraid I'd go through life never giving myself the opportunity to become a writer.

On Monday, March 23, 1998, I flew home from Alaska's oilfield for the last time.

. .

THE PLAN FOR the first week after leaving Alyeska called for Dad and me to fly to Los Angeles, where we met up with Jennifer and Craig, Jennifer's boyfriend at the time and now her husband. The four of us drove to Las Vegas and joined other family members assembled for Emily's marriage to Joe that Saturday.

Mom had been visiting with my sister Risa in Arizona. When the two of them arrived in Las Vegas on Friday night she was experiencing excruciating pain in her abdomen. We took her to the emergency room of the hospital nearest the hotel, and she was admitted. On Saturday, while last-minute preparations were under way for Emily and Joe's evening wedding, Mom was undergoing tests to try to find the cause of her pain. Saturday night she was still in the hospital, still horrifically uncomfortable, and unable to attend the wedding.

On Sunday morning, the newlyweds and other members of the family visited Mom before catching flights out of Las Vegas. More tests had been scheduled for Monday. Risa and I spent Sunday afternoon rearranging our travel and hotel accommodations and settling in for however long we needed to remain in Las Vegas to be with Mom. Sunday evening we went to the hospital to visit her and were shocked by her skeletal appearance. Not to worry, we were told. They were getting her ready for the morning. We should go to the hotel and get some rest.

"It'll be OK," Mom said as Risa and I left, giving her the privacy she wanted while medications emptied her digestive system.

At 10 that evening, feeling uneasy, we called the hospital and were assured preparations for the tests were proceeding as planned. Less than half an hour later someone from the hospital called to tell us they were administering CPR. Risa and I raced to the hospital but were too late. Mom died before we arrived.

Stunned, we called the rest of the family and returned to the hotel to wait while they retraced their steps and came back to Las Vegas. An autopsy indicated that gangrene, caused by a blood clot in Mom's small intestine, had brought about her death.

We were numb with shock. Cast adrift in the universe is how a therapist later described those feelings to me. When I returned to Ninilchik days later, it was to the shelter of the cabin and the embrace of the homestead.

Gee

"5,000 pounds of sandbags"

TRAVELERS INTO HOMER ON THE STERLING HIGHWAY, as it descends Baycrest Hill on the way to its end at the tip of a four-mile finger of land known as the Spit, can see only the top of a hillside-hugging neighborhood named Baycrest Subdivision. Turn right onto Augustine Drive as it branches off from the highway, follow its switchbacks as it snakes downhill through this scattering of homes, and you'll come to Judy Rebecca Court and the gate marking the entrance to Gee Denton's piece of paradise.

On every visit I have been astounded by the beauty her green thumb has brought forth, with spectacular views of Kachemak Bay and the Kenai Mountains as a backdrop. In Gee's carefully tended yard are more than 100 perennials, shrubs, and trees, including her favorite, the Japanese maple.

"I am inspired by how incredible all of God's creation is—so unique and so varied," said Gee, who came to Homer from Colorado in 1999. "I never tire of watching life come after the winter. It keeps life going inside of me."

Gracefully setting aside compliments on her gardening ability, Gee said she views that work as an expression of her faith.

"The Lord sometimes says to me, 'I will meet you in the garden in the cool of the evening' and he does," she said, smiling.

Gee's talent isn't confined to the garden. As we entered the house, her artistry was evident wherever I looked. Found objects expertly placed drew the eye. A collection of doorknobs had become a subtle part of the deck railing. On the kitchen counter was a set of glass dinnerware she made, using the kiln in a friend's studio. Dominating one corner of the living room was a large birdcage she had constructed. In the sunroom sat a table in the making, small pieces of glass being fitted together to catch the light.

In a former life, with a different owner, Gee's home was a rugged summer cabin with an outhouse. Nothing remarkable aside from the view. Now it exudes a calming energy Gee frequently shares with others, whether someone experiencing spiritual turmoil, a friend facing the challenges of day-to-day living, or simply a nosy writer asking questions.

"The most beautiful aspects of my home and garden are that they bring peace and quiet to a noisy harsh world, and that many come to find healing and hope in the quiet place," she said. "It *is* a healing place, a retreat, a place to dream, a place of my heart. A gift. I was given a gift."

. .

IN 2014, SHORTLY after ENSTAR Natural Gas dug a trench in the dirt street below Gee's home and laid pipe as part of a project undertaken jointly by ENSTAR and the city of Homer to bring natural gas to the area, Gee's "gift" began to slip away.

Gee and her Baycrest Subdivision neighbors knew about the natural gas project. She had researched it online. She also had watched ENSTAR videos indicating a sense of stewardship for the environment and the company's plan to remove as few trees as possible. That appealed to Gee, even though she had no intention of signing up for gas.

"It was because of the financial end of it," she said of that decision. "I live by myself, so it was a matter of how long would it take to recoup the costs." Funding from the state would bring the pipeline from gas wells northeast of Homer to the city limits. Paying for a distribution system within the city, approved by city voters, would be done through a special assessment district, with each parcel within the district's boundaries to be assessed an estimated $3,250. Those choosing to take advantage of the new energy source would pay an additional hook-up fee, the amount depending on their distance from the distribution line. For Gee, who owns two parcels—the one she lives on and a neighboring piece she purchased as an investment—managing two assessments was enough to contemplate.

Installation of the gas line in the neighborhood was scheduled for sometime between June and late summer, Gee was told. The timing was good—it would avoid spring, or what Alaskans refer to as "breakup," the time of year when snow and ice melt and frozen ground turns into mud before it dries in the warmth of summer.

"We also were under the impression that we'd be notified when they were going to do it," she said.

Instead, Gee returned home one April afternoon, shocked to discover that ENSTAR's contractor Clark Management had installed the pipeline in the 30-foot right of way on the uphill side of Tanja Court, a seldom-used one-lane road directly below Gee's home. First, the right of way had been clear-cut, denuding it of all vegetation. Then a trench was dug in the still-wet earth, the pipeline installed, the trench filled back in, and the bare right of way spread with wood chips, netting and grass seed.

In May, the front windows cracked in the home of Gee's neighbors Scott and Carolyn VanZant, who live at the end of Judy Rebecca Court. Their toilet no longer sat level, doors wouldn't close, and walking across their deck caused the entire house to shake with earthquake-like tremors. The VanZants discovered leaking plumbing under their home, and the Sonotube concrete columns upon which their house sat had begun to tip downhill.[1]

The parcel on which Gee's house and a second structure, a small log cabin, are located is separated from the VanZants by the property she purchased as an investment. In May, loud snapping sounds in her house began waking Gee during the night. The Sonotubes beneath the log cabin began to tip downhill, just as was happening beneath the VanZants' home, and the cabin began slipping off the foundation. Gee contacted Ben Harness and Dave Northrup of Techno Metal Post Alaska for advice and her fears were confirmed.

"Her cabin was basically falling over down the hill," Dave said when I interviewed him for the *Homer News* after I was made aware of the situation later that year. [2] In three weeks' time, the cabin was raised, a steel pile foundation was installed and the cabin's downhill movement was halted.

But that wasn't the end of it. Water in excess of 100 gallons a day began gushing from the hillside directly behind Gee's home, turning the soil bordering the back of the house into a quagmire. Gee knows the rate of the water's flow because she attempted to remove it a bucket at a time, finally resorting to a pump to move the water a distance away from her home.

Two years earlier, in 2012, Gee also had problems with water after city snow-removal crews melted ice on the uphill side of Judy Rebecca Court and channeled it across the street to just above Gee's home where, during subsequent freeze-thaw cycles, the melting ice flowed downhill onto her property.

"My propane tank heaved over on its side. The back deck heaved a lot, and the skirting at the back of the house heaved, too, and windows started to fail—not crack, but the seals failed," she said. "I called the city and they said, 'What do you want us to do about it?' They were very unconcerned. It was just beyond me, that lack of responsibility in the decision-making and the weight of what that decision does to people's lives."

It seemed clear to Gee that ENSTAR's springtime trenching and clear-cutting in April 2014 had pulled the plug on the groundwater, resulting in the damage she was seeing.

In May, with water pouring out of the hillside and the soil around her cabin turning into a thick soup, Gee contacted then-Homer City Manager Walt Wrede. She emailed photos of the work done to install the gas line and outlined damage that had occurred since it had been installed. Wrede said he would pass her concerns to ENSTAR.

"I don't want to get your hopes up, though," Wrede said in an email to Gee. "ENSTAR had the right to do the work they did within the (right of way)." Having seen pictures of ENSTAR's work on Tanja Court and even visiting the site, Wrede added that although he was not an engineer or hydrologist, "it would be very difficult to demonstrate or prove that this water problem you are having around your foundation" was the direct result of EN-STAR's work.

There was little chance for Gee to get her hopes up. She spent the summer focused on stabilizing the hillside behind her home. She called no fewer than 15 contractors to help with the effort, but it was more than a month before she got any response. One contractor said he'd provide her a written estimate, but that estimate never came. Another contractor who failed to show up was serving on the city council at the time. When Gee later ran into him at a council meeting, he asked, "So, how's it going?"

"I said, 'How do you think it might be going? Because of you not showing up I've hauled 5,000 pounds of sandbags down the hill," said Gee, having singlehandedly stacked the sandbags to create a retaining wall, covered the bags with 80,000 pounds of rock, and hand-dug a ditch to divert the flow of water.

The VanZants also struggled to regain stability for their home. On a visit to them I watched as Techno Metal Post Alaska inserted steel pipes 14 feet into the hillside above the house at a 30-degree angle and constructed a cabling system to attach the house to the pipes. On the downhill side they enforced the foundation with additional piles and had Homer Roofing add reinforced skirting.

After hearing what was happening, John Bishop of Bishop Engineering voluntarily examined the soil and subsurface water for Gee in an effort to assess existing risks. Asked if it was possible to make a direct connection between installation of the natural gas line and the erosion taking place, Bishop said, "There are situations where trenching and the material they put in trenches afterwards can create a way for water to move that wasn't there before."[3]

John also saw a much larger problem.

"That whole hillside is moving. So it would be better if more in the community were doing something, one big effort," he suggested.

Geologist Mike McCarthy also took an interest in what was happening on Gee's property, as well as the VanZants' and in the Baycrest Subdivision generally. He initially identified two possible causes, beginning with the trenching and clear-cutting. The other was water from ditches and culverts that had been neglected and become plugged.

Walking through the neighborhood, it was easy to see what Mike meant by the latter possibility. In many places, ditches were filled with gravel graded by city maintenance crews off the road surfaces. In one instance, John Bishop and I found the opening to a culvert on the uphill side of Judy Rebecca Court but, search as we might, were unable to find an opening for it on the downhill side. In spite of that, when I contacted Carey Meyer, the city's Public Works director, he maintained that the city adheres to a ditch-cleaning schedule.

"When all this came up, it kind of triggered us to say let's make sure all of our drainages are working," Carey said when I interviewed him for the *Homer News* coverage. He went on to say his crew had assured him that "everything is open and it couldn't be unmaintained drainage facilities that could cause anything that anyone is complaining about."

ENSTAR's work was done under a city work permit, according to Carey Meyer.

"We're responsible in some way for controlling the work but did not see a connection between what ENSTAR did and complaints from property owners that damage that occurred on their property was caused by the ENSTAR contractor," Carey said.

As an alternative to trenching, ENSTAR could have bored, a less disruptive method for installing pipe. When I asked Carey about the trenching done along Tanja Court, he said deciding whether to trench or bore had been "group decisions" made by ENSTAR, the contractor and the city.

· ·

GEE AND THE VanZants were not the only residents to suffer damage following installation of the pipe. Identifying the trenching and clear-cutting as the reason a mature birch tree had begun to topple, Shirley Thompson, who lives on the other side of Gee, resorted to cables to help secure the tree. Bryan Zak, a member of the Homer City Council, returned home from vacation, surprised by the amount of clear-cutting along his Rene Court property, two streets uphill from Gee. Bryan also took exception to the lack of clean-up after the work was done. He called ENSTAR to request that it be addressed and was told that it would be. When nothing happened he began posting online videos of the mess along his right of way.

Mary Hogan, who owns four parcels downhill from Gee, was attending school out of state when she received a message around

Easter, late April 2014, that her driveway was a mess. She immediately phoned the city and was told to call ENSTAR.

"I begged him please, not in the spring! I was livid they would dig during spring breakup," said Mary, recalling the phone conversation she had with an ENSTAR engineer whose name she couldn't recall. "He said he agreed and would make sure he talked to the people working with the contractor. He really did reassure me that my driveway would be usable again when they were done, said he had been down there and that ENSTAR shouldn't be working there."

In the summer of 2015 when Mary finally returned to her property she found her driveway and the surrounding area saturated with water and her driveway impassable.

"They brought in equipment at the worst time of the year. Anybody with a brain knows that," she said. "My driveway was destroyed."

In addition, someone had left a used, mangled culvert on her driveway. Mary complained to both the city and ENSTAR about it, asking to have it removed. Neither the city nor ENSTAR would admit having left it, but someone removed it after Mary talked with ENSTAR.

. .

AFTER HER OWN communication with the city manager in 2014, Gee also contacted ENSTAR, waiting a week after the pipeline was installed, giving herself time to be less emotional before she stopped by the company's Homer office.

"I'd been crying a lot and I didn't want to operate out of that place," she said.

After she expressed her concerns and showed photos of what the area looked like, an ENSTAR employee said he would go look at it firsthand. When he called her back, Gee remembers him saying, "I hate to acknowledge this, but it's an awful disaster." He also said he'd see what he could do to get some kind of resolution.

Weeks later, when Gee hadn't received any sort of response, she stopped by the office again.

"He was eating a banana, leaning on a desk, shooting the breeze and finally said, 'Hi.' I said, 'Do you remember me? I left you a few phone messages.' He said he'd been really busy," Gee recalled. She asked what he'd found out about erosion control for the clear-cut area and "he basically said the sum total of the mitigation was to plant grass."

Gee also attempted to make contact with John Sims, then-manager of corporate communications and customer service in ENSTAR's Anchorage office.

"I wrote a lot of letters. To ENSTAR. To the Army Corps of Engineers. I wrote to several departments with the Environmental Protection Agency. Everybody either didn't respond or referred me elsewhere," she said.

For their part, ENSTAR maintained it was without blame.

"From ENSTAR's perspective, it would be a stretch to link any sort of ENSTAR construction-related activities to personal or private property damage," John Sims said in an interview when I was covering the story for the *Homer News*. "There's definitely some damage that occurred at that location, but not caused by ENSTAR construction work. ENSTAR's perspective is we did what the city told us to do."

The city's and ENSTAR's lack of caution with the pipeline project in the Baycrest Subdivision was a surprise to Gee and her neighbors, considering similar history in the area. In August 2006, Homer Electric Association was putting together a plan to provide power to Country Club Estates, a large-lot subdivision located downhill and west of Baycrest Subdivision with its primary access from the beach. HEA's plan of setting poles and stringing line through Baycrest Subdivision required heavy equipment to do the work and the removal of vegetation. Baycrest residents objected to the plan, worried that removal of vegetation would increase damage to the already eroding hillside. In one of several meetings held to discuss the project, Mark Kinney, district conservationist for the Natural Resources Conservation Service, U.S. Department of Agriculture, advised that alders along Judy Rebecca Court should be cut no less than four feet high, leaving roots intact.

"These are not stable soils there," Kinney said. "You have to be careful how you meddle with them."[4]

Having looked at HEA's plan and been told by HEA there would be no erosion issues, City Manager Wrede maintained there was no cause for concern. Then, on August 31, the city revoked HEA's work permit. The reason, according to Wrede, was that extending electrical power to Country Club Estates would encourage more development in the hard-to-reach area. In addition to the beach access, a driveway had been added from Baycrest Subdivision, but Wrede said the driveway was unsafe, could cause erosion to adjacent properties and posed a challenge to providing fire and emergency medical services to the area.[5]

In his letter to HEA revoking the work order, Wrede also said that "the city has concluded that it is not in the public interest to jeopardize properties in an older established neighborhood and risk causing additional erosion and slope failure to their detriment in order to provide electric service to an area which has not seen, but should be subject to normal planning and regulatory reviews before further subdividing and development can occur."[6]

Fourteen months later, HEA developed an alternate route for delivering electrical power to Country Club Estates. However, the utility also argued in court against the city's revocation of the work order. In May 2008, Kenai Superior Court Judge Carl Bauman issued a split decision, ruling that the city had improperly revoked the right of way permit and restoring the permit with the period of construction modified but also finding for the city, denying HEA's claim for potential damages.

That long and well-publicized battle gave Gee and her neighbors reason to believe their environmentally sensitive hillside subdivision would be protected by the city. It would seem that that has not been the case, whether from lack of attention to ditches and culverts, methods for snow removal, or installation of the gas line.

The current situation, whatever inattention or mismanagement may have caused it, is even more surprising when the erosion problems that have plagued the Homer area are considered. Slides along the bluff bordering the bay happen with some frequency. Homes that once sat well back from the bluff's edge now claim narrow safety margins. In 2001, the late Joe Lawlor and his wife, Pat, were awakened during the night by what they thought was an earthquake, only to discover in the morning that their deck and 25 feet of their bluff property had disappeared to the shoreline below. In 2002, when Ocean Drive residents and the city were considering construction of a seawall, one former resident of that neighborhood commented about sea-driven erosion, "You can only slow it down. You can't stop it."

Other areas of the city are built on top of an old lakebed. "People forget about what this all is. You put a road on top, put a bunch of houses around and then forget what it was. . . . What it was is really important when you're talking about bluff erosion," said geophysicist Geoff Coble of Coble Geophysical Services in Homer.[7]

In 2004, the city reviewed its regulations concerning steep-slope development. Summarized by Planning Director Rick Abboud, the steeper the slope, the stricter the revised regulations.

"When the slope is 15-30 percent, you can't exceed development of 25 percent of the lot. Then for 30-40 percent it's 10 percent development. Anything greater than 45 percent needs an engineer to tell us it's not dangerous," said Abboud.[8] With the slope of Gee's property at 30 degrees, according to engineer John Bishop, work wouldn't have required an engineer's approval.

But slope isn't the only factor influencing erosion, according to Geoff Coble. Consider the weight of 2,000 gallons of water delivered to the cistern at a private residence, for example. Or what happens when that water is used and goes into a drain field possibly located near the bluff. Or when vegetation such as the spruce, birch and cottonwood trees or stands of alder common to the Homer area that draw water from the soil are scraped away.

"Transpiration from plants that are that large does have an impact on the hydrology," Geoff Coble said.

For Baycrest Subdivision, another factor in the erosion equation may be the type of drainage system installed by the state Department of Transportation and Public Facilities along the stretch of the Sterling Highway above. Mike McCarthy spent several days measuring water coming through the drain and pouring onto the ground above the subdivision at between 8,000 and 14,000 gallons a day.

"That's when it's not raining," said Mike. "That's a lot of water."[9]

Asked about the drain, Jill Reese, DOTPF's media contact, made clear the specific boundaries of the department's responsibility.

"Doing as much as we can to preserve the road, that's our duty. Preserve the road and keep people moving. That's what we're focusing on," she said.[10]

. .

MOVING ONTO her piece of land didn't come easy for Gee. She first arrived in Homer in 1997, intending to buy a place and stay. Things didn't work out as she planned. Returning to Colorado, she dealt with severe health problems that required hospitalization. She finally came back to Homer two years later and, one thing leading to another, found the place that through much hard work has become home.

Two months after moving in, Gee also bought the lot next door. Shortly after that, one of her neighbors stopped by to tell her not to be alarmed if she heard chainsaws the next day because the neighbor was planning to do some clearing on her land. So, when she heard the noise, she wasn't concerned—until she realized that the trees she saw falling were on her second lot and the cutting was not being done by her neighbor.

"I ran outside and screamed at the guy, 'I don't know what you think you're doing.' He said they were from Homer Electric and were 'going to cut some trees,'" Gee recalled.

She informed the workers that the 50 trees they had cut weren't on an easement, they were on her lot line. The workers left and Gee called HEA, reported the trees having been cut and requested that someone come back and clean up the mess. When no one showed up, Gee followed up with a letter. It was a year before she was contacted and told someone would be out to take care of it. Asked if they could spread wood chips over the area, Gee said no. When she got home that night, the lot was not only covered with wood chips, but a vehicle that had been used had clearly become stuck in the soft soil and left a mess. So, Gee called HEA again.

A representative for the utility co-op showed up, saw the mess and apologized. Gee asked what the plan was to compensate for the lost trees and was told that wasn't possible. When she insisted, she was told he'd see what he could do. "What he could do" became apparent a month or two later, when she received a call from a local garden supply store saying that they had a gift certificate for her from HEA for $125.

In spite of what she's experienced, Gee has not reached the point where she is ready to give up and walk away. With every plant cultivated, sandbag placed, bucket of water dumped, with every artistic creation added, it has become more and more her home.

Of the most recent disruption in her life, which she is certain was set off by the pipeline, Gee said, "It tried to steal my life and my peace and to take my hope away. It has felt like holding on to hope when there was no evidence of hope. It has felt like I have no face. Like all that matters has nothing to do with humanity and everything to do with greed. It has felt like being mocked and pressed down. It has tried to steal who I am and it has been a long and tiring journey. . . . It feels like they *tried*—but they cannot take what is inside me. They cannot take my faith."

· ·

THERE HAS BEEN an occasional spark of hope. During the summer of 2015, friends who had never been to Gee's home paid a visit and were astonished as I had been to see the beauty Gee had created both outdoors and in. Knowing the battle she had waged to minimize erosion and the numerous fronts on which she'd fought that war, the friends offered to help Gee get legal advice. Encouraged, she flew to Anchorage and met with the recommended attorney who, in turn, put Gee in contact with another lawyer she was told would be perfect for her "solid case." That spark of hope was snuffed out, however, when the second lawyer declined because of a conflict of interest with regard to ENSTAR, and Gee was told to expect that response repeatedly should she attempt to pursue legal action.

Gee's neighbor Mary has continued to make calls to the city and to ENSTAR.

"I'm not asking for a new driveway, just a couple loads of gravel to put it back into use," she said. "It's a full-time job just to get the driveway usable."

Considering the damage Gee and others in the neighborhood have sustained since the gas line went in, Mary considered herself "lucky." But the insult of having to pay the gas line assessment on top of the damage done to her land feels like a "violation," she said. "It makes you sick. And there's no recourse."

A year after the gas line was installed, Gee continued to deal with mud. Her house continued to make loud snaps and groans that woke her in the night and made her wonder if the home she has created is losing its hillside foothold. And she wondered about the battle still to be waged to make someone acknowledge liability.

"They need in some fashion to be responsible for the damage," Gee said. "Given the magnitude of what's happened, they need to figure out some form of erosion control. The powers that be need to educate the people doing the work. They need to be conscious of the world around them. . . . ENSTAR, the city of Homer, Homer Electric Association, the Alaska Department of Transportation. Somebody has to be responsible."

Johnny and Viola

"Every year we get that rental, so it helps"

O N A COLD JANUARY AFTERNOON IN 2014 I carefully wound my car over a rough gravel road covered with ice, which led from the Sterling Highway through a maze of gravel pits to the home of Johnny and Viola Hansen. Signs blocking tire tracks that splintered off in various directions kept me on course. No Trespassing. Private Property—Keep Out. Finally, the road penetrated a wall of spruce trees and there before me was the house, spread out along the bluff, facing the wintry gray waters of Cook Inlet.

It was the first time I'd been to the Hansens' home. This home, that is, but not the first time I'd met Johnny and Viola. Viola is a descendant of one of Ninilchik's founding families. Her great-great-grandparents, Grigorii Kvasnikoff and Mavra Rastorguev, came to Ninilchik in 1847. Their son Ivan married Alexandra Sorokovikoff; Ivan and Alexandra's daughter Irene married James Robert Kelly; and Irene and James' daughter Martha and her husband, Torvald Jensen, a Norwegian fisherman who somehow found his way to Ninilchik, are Viola's parents. That may seem an unnecessarily lengthy introduction, but since Viola's great-great-grandparents are my great-great-great-grandparents, it's the long way of saying Viola and I are related, an important detail to members of Ninilchik's first families. Villagers shorten it by referring to everyone as "cousin," but that tends to raise outsiders' eyebrows.

It wasn't the family connection that brought me to the Hansens' that January afternoon, however. It was my curiosity about the drilling rig that towered above the tops of trees and the natural gas flare lighting up the night sky that has from time to time marked a corner of their property near Stariski Creek, on the southern Kenai Peninsula about halfway between Ninilchik and Homer.

Viola was born in Seldovia in the early 1930s, before construction of the Sterling Highway and the access the highway provided between peninsula communities and the world beyond. Just off the Gulf of Alaska's often-stormy waters and tucked inside the entrance to Cook Inlet, Seldovia was a much more active community than it has since become. The completed highway's southern terminus was Homer, a city with about 5,000 residents who pride themselves on living "where the road ends." Separated from Homer by Kachemak Bay and reachable only by water or air, Seldovia is 25 miles beyond the reach of the highway. But in the 1930s, it was the place to be.

"It was where the mail came in for the inlet. And the groceries. And everybody went there at the end of the fishing season," Viola recalls.

In the summer her family traveled from Seldovia, across Kachemak Bay and up the peninsula to their fish camp near the stretch of beach that her home with Johnny now overlooks. Whenever demands of the fishing season allowed, Viola, her brother, Thor, and their mother would make their way another 20 miles or so up the beach to Ninilchik to visit Martha's relatives, either walking or catching a ride on a tender collecting fish from other fish camps along the inlet's eastern shoreline.

"It was so magical. The hills, the river, the beaches. Through my eyes, Ninilchik was a beautiful, beautiful village. The (Russian Orthodox) church was like a sentinel up on the hill. There were milk cows wandering through the town," Viola said.

Besides the physical beauty of the Ninilchik area, Viola's visits to her mother's hometown were an opportunity to enjoy the closeness family ties can create.

"After supper dishes were done, we'd go walk out on the beach, sit on a log and, you know how kids are, especially girls, you've always got something special to giggle about, to talk about," she said.

Seldovia was a long way from the rest of the world, but before the highway went in, Ninilchik was even more remote. Every box, every tin can, every scrap of fabric, every piece of paper was used and then used again. A subsistence lifestyle meant growing, hunting or catching your own food. The language spoken by Martha's generation was a mixture of English and the Russian language of the village's founders.

"I can remember it as if it was yesterday, how they'd all be sitting around a big tub cleaning clams and all be talking in Russian. Mom talked Russian right up until the end. I could understand some of it, but I never was able to converse in it," Viola said. "When I think about it, when I get out and walk now, that Ninilchik beach is my favorite place to go. I think it's because of so many memories. . . . Everything just comes flooding back. Hearing the seagulls, the waves on the beach, even if the water is just slapping on the beach. It's a very special place."

Each fall, with the end of fishing season, the Jensens returned to Seldovia and Viola and Thor started a new school year. That familiar routine was interrupted by the Japanese bombing of Dutch Harbor in June 1942, six months after the attack on Pearl Harbor. With less than a thousand miles separating the Aleutian Islands community of Dutch Harbor from the Kenai Peninsula, Viola's father decided to move his family to Washington state. They rented out their Seldovia home and Viola and Thor were introduced to life in the more heavily populated areas around Seattle.

"It was culture shock for sure. I dressed different and had to learn the social skills needed to survive in school with 1,500 kids when the total amount of kids was maybe 150 in 12 grades in Seldovia. It was just really major lifestyle changes. It all took some getting used to," she said.

The following summer, the family was back at the fish camp, but they continued to divide their time between Alaska and Washington for the next two years.

In 1957, in her mid-twenties and the single mother of two young sons, Viola moved to Ninilchik, the village that held such fond memories for her and where her parents had since settled.

"My mother said just come on up and work at the cannery, which I did, and she babysat the kids," Viola said.

The recently completed Sterling Highway made driving from town to town possible, but without a vehicle Viola and her sons stayed close to home. On Viola's days off from the cannery, however, a friend occasionally picked her and the boys up and took them out for a drive. Telling me the story, Viola remembered that "it was one of my dear friends. Oh, I think it was your mom, such a wonderful lady," a memory that gave us, Viola and me, one of those "we share some history, don't we!" laughs.

It was during one of those drives with my mother that Viola met Johnny.

Originally from Boston, Johnny came to Alaska in 1948 and took advantage of the opportunity to homestead.

"When I went to the land office, the girl there said, 'Where do you want the homestead?' I said somewhere between Homer and Ninilchik and she said, 'Take your pick. It's all open.'"

Johnny finally narrowed his area of interest to 160 acres along the Sterling Highway that spanned Stariski Creek and reached to the bluff overlooking the inlet.

"I'm glad I settled here. After all, this is where they put the pad, so it is worth quite a bit," he said of the exploration pad that has been used by companies looking for oil and gas.

After Johnny and Viola's marriage, the homestead became an active home base for a growing family. To begin there were the two of them plus Viola's two sons, Rocky and Scott. In 1958, Johnny and Viola's son, Torvald, was born, followed by the births of four daughters—Emma, Shelley, Lisa and Heidi—between 1959 and 1967.

Beginning in 1967, according to the State of Alaska Recorder's Office, Johnny and Viola entered into lease agreements with several oil and gas companies. A 1967 lease with Standard Oil Co. of California was picked up by Chevron U.S.A. Inc. and Atlantic Richfield Co. in 1977. Then came an oil and gas lease agreement with Cities Service Co. during the late 1970s and early 1980s.

Oil and gas activities weren't new on the Kenai Peninsula or elsewhere in the state. Drilling for oil in the Swanson River field on the central peninsula had begun twenty years earlier. Farther north, oil had been discovered in Prudhoe Bay in 1968 and production was finally scheduled to begin in June of 1977. However, the Hansens' lease with Cities Service came just months before the tragic deaths of sons Rocky and Scott in a fire in Ninilchik. For comfort then, and for years after, Viola's focus wasn't on what was happening outside her family. She found the healing solitude she needed in what was familiar: fishing.

"I probably fished every daylight hour for 20 years, either on the beach or on Stariski Creek," she said, so she wasn't that much aware of the oil and gas companies and their activity on the homestead land. The interactions she did have with them were generally irritating and unwelcome, coming when Viola would rather be fishing.

"It seemed like that's when the oil people were here, visiting with John. I'd come home and they'd be driving up the driveway. One time they stopped and John must have said I was out fishing and they said they were planning a fishing trip and would I like to go with them, but that wasn't the kind of fishing I did."

Signing their names to an oil and gas lease was one area where Johnny and Viola disagreed.

"You were against it and I was for it, so we did have our little pro and con," Johnny reminded her.

Although she did sign, Viola has had one priority on which she wouldn't bend with regard to the leases: the freedom to maintain her connection to the land on which she and Johnny live, to be able to fish along the creek, walk the beach or through the trees, or simply soak up the soothing sounds of the inlet.

"We use every bit of the homestead. And the beach. And the creek. We have stairs to the beach and I think December is the only month we haven't used them," she said.

Johnny is grateful that the leases augmented his efforts to provide for the family. His work as a heavy equipment operator, including developing geophysical trails in the nearby Caribou Hills, also helped make ends meet but eventually that work ran out.

The leases "supported my family for a good many years. When they started paying, by gosh, that was OK with me," he said. "I think it's been pretty good."

In 1980, according to documents on file with the state of Alaska Recorder's Office, the Hansens signed a 10-year oil and gas lease with Texaco for the use of 144 acres of their land. The lease included a formula for paying the Hansens royalties dependent upon the amount of oil produced, as well as an annual per-acre rental fee. In 2000, they signed a lease with Phillips Alaska, Inc. for 147 acres. It included a bonus and again an annual per-acre rental fee. In 2001, the name on the lease was changed to "Conoco-Phillips Alaska, Inc., formerly Phillips Alaska, Inc."

A chronology of oil and gas development in the area fills in the details of activity on the Hansens' property. It also illustrates the growing interest leading to a project that got under way a year after my 2014 visit with the Hansens: BlueCrest Energy Inc.'s development of an offshore oilfield with drilling to be done from the pad on the Hansens' property.[1]

1967   Pennzoil discovers oil when drilling the Starichkof State No. 1 well with the use of an offshore jack-up rig. The well tests at 75 barrels of oil per day, is plugged and abandoned.

1994   ARCO reviews Pennzoil's data and concludes that the first well may have missed the larger field.

2001   Phillips Inc. directionally drills Hansen No. 1, an onshore well on the leased Hansen property, and confirms Pennzoil's discovery as well as a productive, deeper formation. The Cosmopolitan Unit is formed for further development and testing of the field.

2003   ConocoPhillips (**after 2002 merger**) drills Hansen No. 1A, a sidetrack out of Hansen No. 1; tests at 1,000 barrels of oil a day, produces 14,851 barrels from the well.

2005   ConocoPhillips and Pioneer Natural Gas conduct a 3-D seismic survey, confirming the presence of a larger anticline structure.

2007   Pioneer acquires a 100 percent working interest.

2007   Pioneer drills Hansen 1AL1, another sidetrack, which produces 35,000 barrels of oil.

2010   Pioneer returns to fracture stimulate ("frack") the second well, which flow-tests 250 barrels per day of oil and produces more than 33,000 barrels under Pioneer's operation.

2011   Pioneer disbands the Cosmopolitan Unit but retains two of its leases.

2012 Buccaneer Energy Ltd. and BlueCrest purchase Pioneer's two leases, with BlueCrest controlling 75 percent interest and Buccaneer 25 percent. Apache Corp. acquires the other Cosmopolitan leases in a state lease sale.

2013 Using the jack-up rig Endeavour-Spirit of Independence, BlueCrest and Buccaneer drill Cosmopolitan State No. 1, proving up six new productive gas zones and four new productive oil zones. BlueCrest and Buccaneer acquire Apache's leases.

2014 Buccaneer declares bankruptcy and sells its Cosmopolitan interests to BlueCrest.

. .

IN OCTOBER 2014, the Hansens' 2001 lease with Phillips Alaska, Inc., was amended with BlueCrest Alaska Oil & Gas LLC named the successor. BlueCrest is a privately held Texas-based company focused on acquisition, exploration and development of oil and gas. It has regional offices in Texas and Alaska. "Our company values are centered on fulfilling our commitments to stakeholders, environmental stewardship, and fidelity to conducting our operations with the highest level of ethical standards," says a company handout.

In 2015 the Hansens also granted access to the Alaska Industrial Development and Export Authority (AIDEA), which is lending BlueCrest and its affiliates as much as $30 million to purchase a drill rig, rig camp, and other related equipment and materials for developing BlueCrest's plans for the site. Created as a public corporation by the Alaska Legislature in 1967, AIDEA's mission is to "increase job opportunities and otherwise to encourage the economic growth of the state, including development of its natural resources, through the establishment and expansion of manufacturing, industrial, energy, export, small business, and business enterprises," according to information it provides online.

At a September 2014 meeting in Kodiak of the board of directors of the Cook Inlet Regional Citizens Advisory Council, BlueCrest director, president and CEO J. Benjamin Johnson presented an overview of BlueCrest's Cosmo plans.[2] He reviewed BlueCrest's history partnering with Buccaneer and the 2014 acquisition of Buccaneer's interests and outlined a plan for including the onshore drilling of oil development wells and construction of facilities in 2015. Development of the site would include a 14-foot tall berm to reduce noise and lights from the site for neighbors, as well as provide containment and protection of Stariski Creek and Cook Inlet in case of a spill.

Other plans Johnson outlined for the onshore facilities included installing high-noise equipment inside an insulated building intended to contain the noise. Light deflectors are to be used for lighting at the pad. Grid power will eliminate emissions from the site from power generation, unless the grid fails. Natural gas and diesel generators are to be used for backup power. Containment at truck loading racks is expected to address any spills, which would then flow back into a drain system and be reprocessed and sent back to oil tanks for trucking to customers. Primary containment also will be provided around all tanks "for catastrophic events such as tank rupture. This is an unlikely event, but any rupture/spill is fully contained and will drain to the sump system," Johnson said.

Cosmopolitan's plan called for gas extraction from two offshore platforms. According to the plan as presented, there would be no liquid hydrocarbon production offshore, only dry gas. The system would be computer automated, with multiple redundancies in process safety systems.

"Every component of the process system has been analyzed for potential failures and accommodations made (including procedures) for handling such failure to prevent safety and/or environmental breaches," Johnson said in his report.

A fail-safe design is intended to force a shutdown in the event of system failure.

According to a fact sheet distributed by BlueCrest at several community meetings, oil production of the Cosmopolitan Development Project is set to begin in 2016 and is estimated at 5,000 barrels per day during the first year and as much as 17,000 barrels per day at the end of five years. The life of the field is estimated at 30 years. The oil will be trucked approximately 75 miles north, along the two-lane Sterling Highway to the Tesoro refinery in Nikiski. During peak production, that could mean one truck leaving the site every 45 minutes to an hour.

My attempts to reach BlueCrest's health, safety and environmental manager Larry Burgess by phone during the summer of 2015 were unsuccessful. However, in a January 2016 article by Jenny Neyman of the *Redoubt Reporter*, Burgess basically confirmed the earlier information and said drilling activities on the 37.5-acre pad were anticipated to last five years. He said he thought the drilling rig, being built in Texas for shipment to Alaska, would be "the largest operating rig in the state of Alaska, capable of drilling up to 30,000-foot-long wells."[3]

The total number of employees needed for the project's construction, drilling and operations is anticipated to reach 400, Burgess said in 2015.[4] At that time, he also noted there would be "some good short-term employment for locals." As of January 2016, with the rig still not on site, Burgess said 48 people were employed, about half of them local hires. "We're trying real hard to hire as many locals as possible, but, believe it or not, it's not as easy as you'd think for some of those jobs," he said.[5]

Potential gas production is estimated to be between 60 and 70 million cubic feet per day. The ENSTAR Natural Gas pipeline passes nearby, but BlueCrest's plans for developing the gas assets are still being developed. In the 2016 update in the *Redoubt Reporter*, Burgess said that gas production ". . . is currently on hold, and it will probably continue to be that way until two things change—market and the Alaska tax credit structure. . . . We have a huge gas cap on top of this oil, and we could produce gas from that field for 30 years, probably."

In announcing support of BlueCrest's project, John Springsteen, executive director of Alaska Industrial Development and Export Authority, said, "AIDEA's financial participation in this rig will not only help secure long-term energy supplies for Alaskans, it will create job growth and help advance the region's economy."[6]

· ·

WITH THE PROJECT gaining momentum, I decided in August 2015 to see it for myself. Following the same road I'd used when I visited the Hansens, I soon encountered equipment-hauling flatbed trucks taking up both sides of the narrow road as they left the pad for the Sterling Highway. Pulling over to make sure they had enough room, I returned what I thought was a wave from the driver of the first truck, only to realize he was impatiently attempting to direct my attention to a "No Trespassing" sign. His reaction is a curious one, considering the road does double duty, providing the only access to the Hansens' home as well as the developing pad.

Signs posted on the containment berm surrounding the pad warned trespassers they would be responsible for any damage they did. On the pad itself, pilings were being driven, hard-hatted employees leaning over what were perhaps drawings. Vehicles of varying shapes and sizes hauled supplies from one point to another.

Continuing on to the Hansens' driveway, I was surprised by their home's proximity to the pad. A stand of spruce trees and little else separates them from the mounting buzz of activity. It brought to mind a comment Johnny made during my visit a year earlier.

"We have a full homestead, but they're only renting 30 acres (for the pad)," Johnny said, then adding, as if reminding himself, "Every year we get that rental, so it helps."

Maintaining ownership is important to Johnny. It is something for which he has tenaciously fought in spite of multiple offers.

"Buccaneer was very aggressive. They offered . . . like a half a million for a couple of acres. . . . Two or three different times they came armed to get a yes. . . . Of course we said no way. I vowed I'd die still owning every piece of this place," Johnny said. "When they came and offered us cash, I thought, gee, that would put them in the clear. They'd own the land and we wouldn't have any say. So, no."

"Yeah, they kept coming. Bring flowers. Bringing pies," Viola said.

"That part was all right," Johnny said, smiling. "But they knew the answer wouldn't change."

. .

THE HANSENS have lived in their current house for more than 10 years. It is large and with a basement that provides plenty of space for their family.

"At Christmas, we had 23 here for the holidays. The kids know they have a place to come," said Viola.

Life has eased over the years for the Hansens. For Johnny, the homesteading days that demanded five years of hard work proving up on the property are past. He no longer gets up in the morning facing hour after hour of jarring rides on the heavy equipment he drove and the rough terrain on which he worked. He is rightly proud of his accomplishments, among them the house in which he and Viola now live and of which he is quick to say, "The oil companies didn't have anything to do with the building of this house. It cost us a little over a half million to build it, and that's not counting the land."

Viola no longer has to walk the beach or hitchhike on a passing fish tender or depend on close friends to get out and about. She can fish to her heart's content anytime she chooses. She enjoys reliving the happy memories of her childhood every time she walks the beach in front of her home, remembering her summers fishing with her parents and brother. When her children and grandchildren come to visit, she gladly fills their requests for the traditional salmon dishes her mother taught her to cook.

"These are the things that are going to stick with them and their kids," she said. "It's wonderful that it means so much."

All these years later, would they still agree to the string of oil and gas leases they've signed, I asked them in 2014.

"It's easy to talk about it now, but when you first get into it, who knows? It's just an educated guess," Johnny said.

And what of 2016, if BlueCrest's project continues as planned? If oil-hauling trucks are leaving the pad every hour? If hundreds of workers are reporting for work? If two drilling rigs are parked just offshore? Will Johnny's answer be the same?

Late in the fall of 2015, with BlueCrest's construction on the pad underway, I paid the Hansens yet another visit to ask what they thought of the development taking place.

"As you notice, you open the door and hear lots of noise, but you can't hear too much in here," Johnny said, seated with Viola in front of their living room windows that overlook Cook Inlet.

"I'm just grateful they decided to drill there and not in front," said Viola of BlueCrest's plan to develop the onshore oil wells first. With the date for the offshore gas activity dependent on "gas marketing and investment decisions," according to Johnson's Kodiak presentation, and considering her and Johnny's time of life (Viola is in her early 80s, Johnny in his 90s), Viola added with a soft laugh, "I won't be here to worry about the platforms."

HERFF · JOLAYNE · KEN

"They'll tell you whatever they need to tell you"

LESS THAN HALF A MILE NORTH OF Johnny and Viola Hansen's home is the residence of Herff Keith, reached by Katana Drive, a one-lane rutted gravel road marked by a street sign but more closely resembling a well-worn driveway. The road branches off the Sterling Highway, slices through a wall of alders, descends steeply to Stariski Creek, crosses on a narrow wooden bridge, and makes a final, even steeper, ascent to a small handful of homes.

Avoiding puddles and ruts and following Katana nearly to its end, I pulled into Herff's driveway. On one side, its back to the street, was a large garage. On the other side was a storage building. In the center of a circle drive Alaskan wildflowers bent in a gentle Cook Inlet breeze. Where the driveway circled back on itself was the house, overlooking the inlet from the edge of the bluff, with tall spruce trees to the right and left. More flowers hugged the foundation and hung in baskets from the eaves. Their bright colors contrasted with the pale yellow siding and white trim of the house —in all, a neat and inviting scene, made more so by a friendly welcome from Kiska, Herff's German shepherd companion, and Herff's shouted greeting as he walked toward me from the garage.

Herff led us up onto the porch and through the main entry of his home to the living room's undisturbed view of the inlet.

Designed by a Canadian firm, Herff's house was shipped to this location as individually numbered pieces he and a crew of helpers assembled over a waiting basement like a three-dimensional puzzle. A unique wall construction minimizes moisture. An air exchange system balances indoor stuffiness with outdoor freshness. Separate heating zones maximize individual comfort and minimize heating costs, and cathedral ceilings in the living room and kitchen add to an airy impression of the 2,000 square feet of living space.

"I've put my retirement into this house," Herff, a former engineer with the Alaska Railroad, told me with no small amount of pride as he set a pot of coffee to brew and Kiska curled up in her bed. "$400,000 out of pocket. It's all paid for."

As the aroma of the brewing coffee gradually filled the kitchen, Herff pointed out a collection of agates of varying sizes spread along the windowsill that testified to hours of beachcombing. He showed me a dried sea star and we marveled at the way its legs curled around rocks it appeared to have been embracing when it died.

From a hutch in the dining area, on top of which sat a jar holding a bouquet of feathers, Herff retrieved a large razor clam shell to show me the size of clams he was once able to dig along this stretch of coastline. In February 2015, the Alaska Department of Fish and Game announced the closure of clamming on the inlet's eastern shore, including the beach in front of Herff's home. An abundance study done in 2014 north of Stariski, near Ninilchik, indicated the average number of mature-size razor clams was an estimated 80 percent less than averages between 1991 and 2012. A study even farther north, in the Clam Gulch area, showed an 89 percent decrease between 1989 and 2008.

"The cause of the decline in razor clam abundance on Eastside Beaches is unknown," the state's news release read.[1]

. .

HERFF'S FATHER WAS the first of his family to come to Alaska. He had been serving in the U.S. Marines when Pearl Harbor was bombed on Dec. 7, 1941. When the Japanese focused their attention on the Aleutian Islands, he became part of the U.S. military build-up in that part of the territory.

"Following the war," Herff said, "when my dad's sister moved to Anchorage, Dad packed us kids all up and drove up the Alcan (from Texas) in the late 1950s."

After graduating from high school, Herff served four years in the U.S. Navy and upon his discharge in 1972 went to work for the Alaska Railroad. He retired from the railroad in 2007.

Anchorage had been his home base all those years, but after retiring Herff decided it was time to "get away from the big city." He didn't want to leave Alaska. He simply wanted something less populated and less noisy. More remote and more closely connected to the outdoors.

"I sold my house in Anchorage, which was paid for, and bought this property. I loved it," he said.

What was there not to love? When it came to fishing, he could choose from the Anchor River, Stariski Creek, Deep Creek and Ninilchik River. For groceries and supplies he had only to go to Homer or Soldotna, no more than an hour's drive to the south or north. If he couldn't find what he needed that close to home, he could always make the 200-mile drive to Anchorage, "but anybody that goes to Anchorage from around here says they can't wait to come home," he said, and he counts himself among that group.

When he bought his land, Herff knew a wellhead existed on the Hansens' property.

"But they weren't doing anything with the well, and I assumed it would be like other wells near Ninilchik—quiet," he said.

Unbeknownst to Herff, the Stariski area's history of oil and gas exploration has been anything but quiet. It dates back to at least the 1960s, although the work has not been constant. It has peaked and declined in rhythm with the market and advancements in technology. After Herff moved into the neighborhood, activity on the piece of homestead the Hansens have leased for oil and gas activity over the years took an upward swing.

"They had a rig here for two years and drilled and burned off the natural gas. Can you imagine having burning gas right there?" Herff said. The wellhead sits on a 38-acre gravel pad and is 1,200 feet from the water well he drilled on his property. The northern-most corner of the pad extends nearly to the bluff, about halfway between Herff's and the Hansens' houses.

On days when strong winds rolled into Cook Inlet from the Gulf of Alaska, the flare's bright flame was stretched in whichever direction the wind was blowing. When winds came in gusts, they would set the flare to dancing unpredictably this way and that, its roar a constant background to its movements.

"Then that quit and it's been nice and quiet," Herff said. "But little did I know what the Dallas and Fort Worth big boys were up to."

In 2014, BlueCrest Alaska Oil and Gas LLC made known its plans for onshore oil and offshore natural gas development in the Stariski area, a project dubbed the Cosmopolitan Development Project. Privately held BlueCrest's base of operations is in Fort Worth, Texas, with offices also in Houston and Anchorage. The Cosmopolitan project as described by the company is much more ambitious than simply reusing the existing well.

Plans call for directional drilling from the gravel pad on the Hansens' property to a 2,759-acre offshore state lease. Natural gas development plans call for two offshore platforms and a subsea pipeline connecting to the onshore facility.[2]

"Just think about it," Herff said. "If you look at a map of the reserve out here, it's huge."

As the pad's closest neighbor, he said, he has had visits from Larry Burgess, BlueCrest's health, safety and environment manager.

"He came here and sat with me, telling me he wants to be good neighbors, but you can't," said Herff, recalling the disruption caused by previous activity at the pad. "When that thing fires up, we will not be able to hold a conversation in the driveway with the noise. . . . When they start drilling, there'll be clanging pipes, back-up alarms, big lights lighting up the sky. We won't be able to look at the stars at night."

During the first part of the summer of 2015, when pilings were being pounded into place on the pad, Herff used an app on his cell phone to track the level of noise. It measured 84 decibels, within the range of a garbage disposal or dishwasher, a freight train at 15 meters, a car wash at 20 feet, a propeller plane flying overhead at 1,000 feet, a diesel truck traveling at 40 miles per hour 50 feet away, a diesel train traveling at 45 miles per hour at 100 feet, with the possibility of causing hearing damage during an eight-hour exposure.[3]

"I texted Larry (Burgess) because I wanted him to know what was going on," Herff said of his effort to impress on BlueCrest the effect their activities are having on the neighborhood.

In a 2016 update on the project, Burgess said, "Now, when you've got a 3,000-hp rig sitting on this pad, some turbines generating electricity, some generators and trucks with backup alarms running 24 hours a day, you're impacting your neighbors. There's nothing you can do about not impact(ing) them, it's what we can do to minimize the impact."[4]

In addition to concerns about how BlueCrest's plans will disrupt his home, Herff is worried about the company's plans for transporting the oil. The stretch of the Sterling Highway that trucks will enter from the pad is a half-mile, no-passing area where the highway makes a curving descent in its approach from the south, crosses the bridge spanning Stariski Creek and then climbs again, making a sharp turn to the left as it continues north. Only two-tenths of a mile separates the bridge from the entrance to the access road leading to the pad.

"They talk about a turn lane, but (the highway's) not long enough to put in a turn lane because the bridge is in the way. It's a bad situation," Herff said.

The possibility of flooding also has him concerned about trucking the oil. In 2002, flooding dislocated a large culvert and took out a section of the Sterling Highway at Stariski Creek and the bridge at Deep Creek, disrupting transportation and isolating residents on the southern peninsula for several days. Groceries were flown into Homer on Northern Air Cargo's DC-6. Residents cut off by the rising water called for help, and emergency responders relied on helicopters to reach them. A month later, rains returned and temporarily shut down the highway for a second time.[5]

Adequate training for the volunteer firefighters and EMTs of the Anchor Point Fire and Emergency Service Area is another worry.

"These volunteer guys are not trained for what a fire would be like," Herff said, imagining the intensity of a fire on the pad and the level of response it would require. "They need to train our volunteer fire department—and I'd love to see them be a 'good neighbor' and buy a fire truck that could handle it. If we can't get anything else, let's get them to help what we have here."

Al Terry, Anchor Point Fire and Emergency Service Area chief, has met with BlueCrest personnel. Chief Terry didn't mention to me any discussions of additional training or new fire trucks, but he said BlueCrest had assured him that in an emergency responders would have direct access through the pad to the homes in Herff's neighborhood.

"We'll be able to go through the BlueCrest site," Chief Terry said. The other option would be using Katana Drive, "but that's a bad mess," he said of the road that connects Herff and his neighbors to the highway.

The chief also has been assured that BlueCrest "doesn't want to disrupt the neighborhood and community, so daily Alaska life can keep on being what it's been."

Herff's interactions with oil and gas personnel have been less reassuring, however, leaving him with the impression that "they don't give a shit. They'll tell you whatever they need to tell you."

With our coffee cups empty, Herff invited me to join him and Kiska on a walk to the pad to see what might be going on. That was something he did every day, he said, just to keep an eye on things. Our short walk was on a path Herff's daily visits had created, through waist-high fireweed, alders and highbush cranberry bushes ready for picking and up to the top of a berm encircling the pad. The14-foot-tall berm is meant to help reduce noise and light and serve as "containment so nothing can drain off the site and especially into Stariski Creek or Cook Inlet," according to Blue-Crest President J. Benjamin Johnson.[6] Climbing the berm's steep side, recently seeded with grass, we stood for several minutes and watched the construction activities under way.

Herff's observations in his neighborhood and from the air in his Cessna 172, a plane he's flown for 28 years, have taught him that development of the Cosmopolitan project is probably a prelude to further development.

"This is just the start of it," he said. "They're stepping on them in ANWR, not letting them drill there, so here they are. In our backyard. . . . When you fly over this area, in the 10 years I've been here, it's crazy all the rigs that are here."

As we walked back to Herff's yard, I asked if he thought he could live with the conflict between the activity taking place within the berm, so close to his home, and the quiet Alaska life he was seeking when he moved to the peninsula. He had already initiated discussions with a realtor, Herff said, about selling the home paid for with his retirement savings.

"He said he could put it on the market right now and this probably wouldn't affect it. But probably the only person that would move here is someone with a year-round job over there (on the pad)."

As oil and gas development continues to spread on the southern Kenai Peninsula, Herff predicted, others who could afford to move away would do exactly what he's considering.

"We all moved down here for the peace and quiet," he said. However, he added, considering the nearby pad and the oil and gas activity gaining momentum, "We're not going to get rid of them, but they're getting rid of us. It's sad to me. Very sad. We're so screwed down here."

JOLAYNE

ON THE OTHER SIDE OF Stariski Creek is the home of Jim and Jolayne Soplanda, a large, two-story log house they built and positioned to take full advantage of the view. It overlooks the creek below and faces the inlet's entrance to the gulf. It also looks toward the Hansen pad.

Jim and Jolayne met in the Talachulitna River area, approximately 51 air miles northwest of Anchorage and accessible by floatplane. After marrying, they lived along the river for 10 years. Jolayne homeschooled their children. As the children got older, she said, she realized they needed more than she could teach them, especially in the areas of math and science. The couple began looking for someplace else to call home.

"We'd always lived in the wilderness, so we knew we didn't want to move to Anchorage. We started watching the news and saw that this area was warmer, so (in 1995) we came here," Jolayne said.

Eager to maintain a wilderness setting, the Soplandas left standing as many trees as possible. That was a decision for which they were thankful when previous oil and gas drilling on the Hansen pad commenced, and bright lights made it possible for operations to continue around the clock.

"We are so happy we left this tree," said Jolayne, pointing to one tall spruce just to the left of the dining room windows. Indicating the second floor overhead, she said, "This is our bedroom, and that tree really blocked the lights that were on all night."

Like Herff, the Soplandas heard the roaring natural gas flare. And they weren't the only ones disturbed by its presence.

"I'm a dispatcher for Anchor Point (Fire and Emergency Service Area)," Jolayne said, "and we'd get 911 calls constantly because of the flare, people thinking there was a fire."

The Soplandas own other property in the area. As with Herff's land, it was bought in quieter times, when no oil and gas activity was taking place. One of their properties is on Katana Drive, near its intersection with the Sterling Highway. Jim has built the family a new home at that location, and their log house is for sale. Unfortunately, the new site "looks right at the pad," said Jolayne. "Jim was so excited when we got that property, but now . . . well, there's not too much we can do about it."

KEN

FOR EACH PERSON eager to share the impact that oil and gas activities on the southern Kenai Peninsula have had on their lives, there also were those who asked that their stories not be told. Some had signed leases and were bound not to discuss details.

Others had contracts to provide services to exploration and production companies and their contractors and, in return, had agreed to keep their dealings confidential.

Ken Lewandowski was one of those who asked that his name not be used when we met in April of 2011. At the *Homer News*, my co-workers and I were closely following the progress being made to bring natural gas to Homer and neighboring Kachemak City. The possibility had been discussed for decades and apparently was about to become a reality. As that effort was taking shape, Ken called me at my desk late one Friday afternoon to introduce himself and strongly encourage me to see the impact of the project at his Anchor Point home. The following Sunday, camera and notepad in hand, I found my way to Ken's log house. My husband Sandy came along, as curious as I at what we might find.

What we found was worth the 15-mile drive. The catch was that Ken, in spite of his anger at nearby activities causing disruptions in his life, refused to have his name used, did not want photos taken, and did not want his address divulged. Living in the midst of a community that seemed overwhelmingly welcoming of the employment opportunities and cheaper energy the project promised, he was wary of repercussions from his criticisms. I respected Ken's wish, but I knew the *Homer News* policy against running stories without a nameable source. I told him that, but he was firm in wanting to remain off the pages of the paper.

In 2013, after hearing more accounts of Kenai Peninsula residents' life-altering encounters with widening oil and gas interests on the peninsula, and having begun work on this book, I reconnected with Ken. Would he consider letting me tell his story? This time, he agreed.

· ·

THE CITY OF Homer, with a population of about 5,000 on the shores of Kachemak Bay, has a long history of attempts to bring

natural gas to the area. For various reasons, one project after another has failed to materialize. In 1988, for instance, Mountain
Alaska Energy was able to secure neither financing nor a gas
supply. In 1995, Alaska Intrastate Gas announced it wanted to
make Homer the hub of a project to supply liquid natural gas
to an area stretching from Ketchikan to Kodiak, but that also
never happened.

It was a plan championed by Rep. Paul Seaton that finally
found traction. Seaton, a commercial fisherman and as noted my
husband's son-in-law, was elected to the state House of Representatives in 2002, his district including the southern Kenai Peninsula.
At a Homer City Council session on Oct. 13, 2008, Seaton made a
pitch for city officials to meet with ENSTAR Natural Gas and
Armstrong Cook Inlet, operator of the natural gas wells on the
North Fork Road, approximately nine miles east of Anchor Point,
a small community about 15 miles north of Homer.[7] He eventually
convinced the council it was in their best interests to help bring
natural gas to Anchor Point, and then by pipeline along the Old
Sterling Highway to Homer. Liking what they heard, the council
added the project to their legislative priority list.

Alaska Gov. Sean Parnell was not so accepting of the idea,
however. In 2010, Homer's legislative grant request for the project
was redlined by the governor. Almost. He left $525,000 on the
table, enough for the city to contract with ENSTAR to build a line
from the well pad to Anchor Point, construct a gas regulation station, and lay about 2,680 feet of pipe that could later be used for
completing the distance to Homer.[8]

ENSTAR Natural Gas is a division of SEMCO Energy, an in-
direct wholly owned subsidiary of AltaGas Ltd. of Canada. For
more than 50 years, ENSTAR, a regulated public utility, has provided natural gas to homes and businesses in Alaska's Southcentral
region, mostly Anchorage and its surrounding area. According to
information provided by ENSTAR, it has approximately 139,000

customers in an area with more than 300,000 people in a state with a population of 736,732.

In 2011, Parnell again gave a thumb's down to the project funding request. However, with the gas regulation station completed and a pipeline being constructed to the north, to deliver gas from the Armstrong wells to ENSTAR's urban customers in the Anchorage area, Homer renewed its push to bring natural gas to the southern end of the Kenai Peninsula.

The next year, Parnell OK'd an $8.1 million grant to build the pipeline from Anchor Point to Homer and Kachemak City. In 2013, the Homer City Council passed Ordinance 13-03(S)(2) authorizing a special assessment bond in an amount not to exceed $12.7 million to finance the design and construction of the natural gas distribution system within the assessment district, with boundaries basically identical to those of the city.[9]

It was in April of 2011, when construction of the gas regulation facility near Anchor Point was undergoing its final tests, that Ken called and urged me to stop by his house.

The gas regulation facility is situated on the corner of North Fork Road and Bailey Drive, the latter the only street in and out of the five-parcel subdivision where Ken lives. What you see to the right as you turn onto Bailey Drive is a fence, topped by three rows of barbed wire, enclosing a single blue structure, a series of gauges, and a network of piping. Posted on the fence are "No Trespassing" signs and others that warn of unodorized natural gas and the presence of a high pressure gas line, and providing ENSTAR contact information. Just inside the subdivision side of the fence, a six-foot-tall wall of gravel and dirt has been created.

The five houses in the subdivision, on lots between one and two acres in size, are separated by original stands of trees and bushes that have purposefully been left in place. The natural foliage and the distance between the homes ensure privacy, something people in this part of the country highly prize. Ken's home is two houses in, on the same side of the street as the gas regulation facility.

Originally from New Jersey, Ken first came to Alaska in 1985 and fell in love with the Anchor Point area. Like many Alaskans, he has found employment in the oil and gas industry on occasion, even working as part of a seismic crew on the Kenai Peninsula. He moved in to the small, quiet subdivision more than 10 years ago.

The gas regulation facility has shattered that peaceful setting. Activity at the facility was causing Ken's house to vibrate with such force that items on shelves and rafters were in danger of falling to the floor. Ken had artfully used fishing line to help keep his belongings in place, but wall-hung pictures were knocked askew. In addition, although it was silent the short time Sandy and I were there, Ken described noise from the facility as so loud as to be deafening. It wasn't constant, but when it started up, its volume drowned out all other sounds.

Ken's attempts to report the disturbance to facility personnel were frustratingly fruitless. Concerned about what was happening and fearful of what might happen next, Ken called the local Anchor Point Volunteer Fire Department.

"They came by and said, 'Oh yeah, it's flaring.' And I said, 'OK, well, why weren't we told? What are they flaring?' And they said, 'Oh, just stuff in the pipe. It's nothing to worry about,'" he recalled.

Ken also questioned the purpose of the facility. According to a set of drawings he managed to get his hands on, the facility was actually identified as a "benzene reduction unit." According to the Oil and Gas Accountability Project, benzene is "an aromatic hydrocarbon present to a minor degree in most crude oils."[10] A highly volatile substance, benzene is described by the Center for Disease Control and Prevention as a natural part of crude oil and gasoline. Long-term health effects from exposure to benzene are on the blood; the U.S. Department of Health and Human Services has determined that benzene causes cancer in humans.[11] The World Health Organization associates benzene with a wide range of both acute and long-term adverse health effects and diseases, the list including cancer and aplastic anemia. Symptoms of benzene exposure include headaches, dizziness, drowsiness, confusion, tremors and loss of consciousness. Alcohol consumption can exacerbate its effects. It also is known to be an eye and skin irritant. Ken feared the risks that the presence of benzene would introduce to his neighborhood. Wanting a first-hand look at the operation, he requested a tour of the facility.

"We got to the gate, opened it and (the company representative) said that's where the tour stopped. He said we didn't have safety gear on. I said if I had known, I would have had it, but he said, 'You don't have it so this is where the tour stops,'" Ken recalled.

From where he was standing, Ken asked what was in each of the pipes he could see. That and subsequent questions went unanswered and Ken went home even more frustrated and suspicious of his new neighbor. A lack of communication also had been evident at community meetings presumably held so local residents could learn more about the project. Ken said he had the distinct impression "nobody wanted to talk to us."

On his own, Ken continued to do his best to piece together what was happening with his new neighbor, desperately wanting to understand the latest activities and how they might affect him and others living nearby.

"They kind of discharged everything into containers. Instead of the atmosphere, it was into containers and they let it dissipate," he said. "They said it was just part of them clearing the pipe, getting everything out of the pipe, and once they cleared and pressure-tested it, they'd be able to put everything back in and get the gas to the facility."

He took exception to the lights that shone into his living room "every night of the week—they never shut them off." When he asked about the presence of the dirt barricade that separated the facility from the neighborhood, he was told it was a "beautification berm, but the only thing I can think of is that it'll hold back the blast if and when something happens," said Ken.

At one community meeting, Ken directed his questions to Charlie Pierce, ENSTAR's division manager on the Kenai Peninsula. In addition to his position with ENSTAR, Pierce was serving on the Kenai Peninsula Borough Assembly. He was elected in 2008 to represent the peninsula's Sterling and Funny River areas, served as vice president of the assembly in 2010–2011, and "termed out" in 2014.

"I remember arguing with him, telling him I thought they were very deceptive in the way they got into our subdivision, but he said they weren't in the five lots of the subdivision, so why was I complaining," Ken said.

When Pierce asked Ken how he could complain about a "new blue house" being added to the street, Ken responded, "Just a blue house, not a blue building with pipes that you guys aren't willing to tell us what's in them?" That brought the conversation to an end, he recalled.

Maybe it wasn't specifically a part of the subdivision, Ken said, but subdivision residents had to cross over the pipe in order to enter or depart their neighborhood. Concerned about that arrangement, Ken took his questions to Kenai Peninsula Borough Mayor Mike Navarre.

"I said what are you going to do about the people of Bailey Subdivision if something happens. There's no way to get us out without having to cross that high-pressure pipe and past the facility, so what are you going to do about it," Ken told Sandy and me. "I called him and called him and called him about it, and he did absolutely nothing to help us out."

Contact with Ken's representative on the Kenai Peninsula Borough Assembly, Mako Haggerty, was equally unsatisfactory.

"He was all for it," said Ken of Haggerty's support of natural gas finally being available to southern Kenai Peninsula residents. "I was really disgusted with him. I didn't find any good representation." (Note: Telephone and email attempts to reach Mako Haggerty were unsuccessful. An interview with Mayor Navarre is included in the next chapter.)

Ken was convinced that in the event of a disaster involving the gas line or the facility, another road providing access in and out of the subdivision was the best solution. He ran that scenario past the assistant chief of Anchor Point's fire department.

"I was told that if there was a problem, we were guaranteed to be out within two hours," said Ken. "He didn't say how he'd get us out of the subdivision, just that he could get us out. I said, 'Alive, or our burnt, charred bodies?'"

· ·

ADDING TO HIS frustration was the lack from the beginning of an ENSTAR offer to provide the neighborhood with natural gas.

"Without a doubt, it should have been considered in our subdivision, but they never offered it to us at all. The only thing I ever heard from Charlie Pierce was if we gave them money, they'd get the gas for us," said Ken, referring to formation of a Kenai Peninsula Borough utility special assessment district (USAD). Once a district is approved by the borough, each property owner pays a portion of the cost of the improvement. It was a cost Ken and his neighbors were unwilling to bear.

Residents in another area of Anchor Point chose to jump through the hoops needed to tie into the new energy source. The Anchor Point USAD, formed in 2011, encompassed 268 parcel owners. For a total of $723,410, they had a main line constructed, with each parcel owner assessed approximately $2,700, payable over 10 years. Individual costs to connect to the line were individually paid.[12]

Residents just north of Anchor Point wanting to tie into natural gas formed another USAD in 2012. It encompassed 49 parcels, each assessed at $3,315, according to the Kenai Peninsula Borough Assessing Department.

"They definitely divided and conquered us, the people who can afford it and the ones that can't," said Ken, maintaining that ENSTAR should have paid for the distribution system within the community. "I just thought they were going to offer it to us. How stupid could I have been? They never even offered it to us. It was never a consideration. That's a tough one to stomach."

Ken was also critical of project advocates referring to natural gas as a cheaper source of energy.

"I've followed the tariffs in the last year and a half and they've had six increases in the last year on natural gas already. I've always said 80–85 percent of people in Anchor Point couldn't afford it even if it was right next to them and was brought into their houses for free," said Ken. "The people that got it were the affluent people, people that have more money than any of us in town."

· ·

ANCHOR POINT'S name dates back to 1778, when Capt. James Cook is said to have sailed into the inlet and lost an anchor in the area.[13] An 1890 census listed three men involved in gold-mining operations along the beach. In the 1940s, homesteaders began moving into this part of the peninsula.[14]

Today, with a population near 2,000, Anchor Point has a K-8 public school with about 100 students, a handful of businesses and churches, a chamber of commerce, the Anchor Point Senior Center, the Veterans of Foreign Wars Virl "Pa" Haga Memorial Post 10221, and an Alaska State Trooper post. A state campground spreads along the banks of Anchor River, where the opportunity to sportfish depends on the Alaska Department of Fish and Game's count of salmon going upstream to spawn. A commercial boat launch facility makes it possible to fish on Cook Inlet from privately owned vessels or with a fishing charter for salmon or halibut. At mile 156 of the Sterling Highway, a sign identifies the site as North America's most westerly highway point. The median household income in Anchor Point is $53,359, according to data collected for 2010–2014 by the American Community Survey, conducted by the U.S. Census Bureau. The survey also indicated that 11.2 percent of Anchor Point individuals were living below the poverty level.[15]

Not wanting to sell his log home, but eager to be away from what was happening in his neighborhood, Ken purchased property about 10 miles north, near Stariski Creek, on Katana Drive.

"And then within two years of buying the land, we had a well by our front door," he said of the oil and gas development on neighboring property owned by Johnny and Viola Hansen. "I can hear the back-up alarms. The rig's right there. The bright lights. . . . Talk about having oil and gas in your backyard."

In 2015, with BlueCrest's project taking shape on the Hansen well pad next to his Stariski property, Ken's plans for the future were uncertain.

"The whole reason I bought the Stariski property was to move to Stariski, but I don't see that being a viable spot at all," he said. "This is a really, really frustrating position to be in. I moved out of New Jersey, the most toxic, Superfunded state in the nation to come to Alaska," said Ken. "I can't believe I left Superfund sites for this. . . . Where is the way to stop these guys? I don't even know where to turn. . . . People need to know what this is like."[16]

LEADERSHIP

"We never look beyond the most immediate emergency"

HUSBANDS, HEALTH AND WEATHER WERE THE expected topics of conversation as Barbara Howard and I enjoyed a rare visit at one of Homer's restaurants in November 2015. In short order, however, the subject shifted abruptly when Barbara asked how work was progressing on this book

Barbara and I had met while she was serving on the Homer City Council and I was a *Homer News* reporter sharing coverage of the council's activities with my co-worker Michael Armstrong. Barbara began serving on the council in 2008 but after what she described as "life altering surgery" decided not to run for re-election in 2014. After I retired from the *Homer News* in February 2015, Barbara and I could enjoy each other's company without her guarding her words lest they end up in print and me able to set my reporter hat aside.

Barbara had heard I was writing a book. In the course of our chatter, around chips, salsa and tostadas, she asked how it was progressing. Then she asked what the topic was. Then she asked how I'd come up with the topic. I explained that in the course of covering the city's plans to bring natural gas to Homer it had seemed appropriate for the newspaper to ask people living nearest the gas wells some 24 miles to the northeast by road how the project was impacting their lives. We had interviewed Mark and Juley McConnell, who lived across the street from the well pad, and Rick and Lori Paulsrud, next-door neighbors to the pad, with 300 feet separating their house from the 24-hour-a-day activity.

The McConnells said they were not bothered by the noise or the lights. Neither were they concerned about increased traffic along North Fork Road on which their home and the pad were located. For them, the work at the pad was a good thing—it had resulted in full-time employment for Mark. So enthused were they over what was happening, the couple had posted a "Drill, Baby, Drill" sign in their front yard.

For the Paulsruds, the light, noise and nonstop activity on the pad had shattered the quiet, close-to-nature environment they had carefully chosen 11 years earlier as the spot where they wanted to raise their son. A realtor had offered to help draft a request to have Armstrong Cook Inlet, the well pad's operator, purchase their house and property so they could relocate. The realtor also told them that if they chose to sell there was little likelihood they could get what they'd invested in their home because of its proximity to the well pad.

Barbara, who as a city council member had helped shape Homer's natural gas distribution project, was stunned by the Paulsruds' story.

"I didn't know we were hurting anyone," she said, staring across the table at me. "All I could think about was getting a new stove."

The look on Barbara's face and surprise in her voice stayed with me over the next few days. Clearly, she hadn't been aware there was a flip side to the city bringing a cleaner, cheaper energy to Homer than the electricity, heating oil, propane, wood and coal already available. Hearing that it had destroyed the home a family had created caught her by surprise.

Several days later, Barbara and I continued our conversation over cups of coffee in her living room.

"We were so behind the curve in having natural gas compared to all of the United States, and the need to have a heating source alternative seemed to be an economic issue as well as a health issue," she said. "And there were all the new appliances that could be bought, and the plumbers having a heyday with work to do   it really was a big boon to businesses." As to the personal economic issue, she said, she and husband Bob "used to burn $600 a month in oil and now we probably burn $200–$300 a year."

Barbara had served as city clerk for Morgan Hill, Calif., before moving to Homer in 2004. She drew parallels to complaints she'd heard from Morgan Hill residents when their scenic views or their farm and grazing lands were negatively impacted by development occurring on neighboring pieces of property.

"We got taken to court many times, and always the judge would rule in favor of the property owner having the right to develop the property to the highest and best use that zoning permitted—and if you didn't like it, then buy (that) land and keep your view. That was my first encounter with development versus people's rights," Barbara said.

Barbara recognized the vision many have of an Alaskan lifestyle, she said, a "dream of living a peaceful life, wanting more animal life than human life around them, tranquility, the hardships and the 'goodships.'" However, she said, "Alaska is evolving to be more like the Lower 48 than most people are probably ready for." The conflict between that Alaska dream and the rush to catch up with other places in the country was especially evident to Barbara at candidate forums where she heard strong anti-zoning and anti-legislation sentiments. Without those safeguards, she said, the

"During the construction of the oil pipeline

in the 1970s and the revenue boom that followed," Alaska economist Neal Fried wrote in 2013, "Alaska's population surged like never before, followed by the state's only economic bust and its largest-ever outflux of population.

"However, even after those losses, Alaska's economy remained larger than it had ever been. The state's workforce recovered quickly and grew every year except one over the past 25 years.

"Today, oil funds over half the state budget—56 percent in fiscal year 2012—and about 90 percent in state general funds."[1]

Fried's article includes a striking assessment from Alaska historian Terrence Cole: "The balance sheet of Alaska history is simple: One Prudhoe Bay is worth more in real dollars than everything that has been dug out, cut down, caught, or killed in Alaska since the beginning of time."

According to a 1985 report on the state's growth and migration written by Greg Williams for the Alaska Department of Labor, in the period from 1973 to 1976, the years the trans-Alaska pipeline was being constructed, the average annual growth rate was 6.6 percent. That slowed during the 1976-1980 period following construction; however, growth continued to outpace the state's out-migration. In 1981, another economic boom took place in Alaska, largely due to an increase in spending at both the state and private level based on oil prices and state oil tax revenues, according to Williams, and the state experienced a 30 percent population growth in a four-and-a-quarter-year period, "making it the most rapidly growing state in America." [2]

In 1959, the year Alaska became a state, its population was 230,500. By the end of 1984, that figure had more than doubled, to 523,048. As of August 2015, the Alaska Department of Labor numbered the state's population at 735,601.   −McKJ very resources we value remain unprotected and subject to destruction "because of our so-called freedom. Self-regulation is a lofty thought," she said.

Barbara's comments were not dissimilar from those I heard expressed by other elected officials representing the southern Kenai Peninsula. Theirs was a perspective extending beyond individual backyards, neighbors and disruption to everyday life. Their vision was far different from what had drawn me back to my piece of the family homestead. The Alaska state song lauds "the blue of the sea, the evening sky, the mountain lakes, and the flow'rs nearby," but then summons "the gold of the early sourdough's dreams, the precious gold of the hills and streams." Had the pristine sea, sky and lakes been eclipsed by the sourdough's dream now turned to the black gold of a growing oil and gas industry, I wondered? Were we turning or had we already turned a corner driven by a growing industry that brought with it a growing population and what that population thought it needed?

Joining our conversation, Barbara's husband, Bob, gave a nod to the state's increasing population and the inevitable changes that would mean.

"You look at people like Daniel Boone. 'There's just too damn many people here. I've got to get out of here.' And so he moved up the Missouri River to get away from people, but the people followed," Bob said. "The pioneers moved in, but then there's the whole horde that follows them."

· ·

ALSO A FORMER Homer City Council member, Beth Wythe was elected mayor in October 2012 and again in 2014. As the city's plan to bring natural gas to the southern Kenai Peninsula proceeded, Beth's employment with Homer Electric Association became a sticking point. The council found that a conflict of interest existed and exempted her from any discussions or decisions on the subject.

The natural gas distribution project included laying 72 miles of pipe so that gas could be delivered to 3,855 city properties. Its $12.7 million price tag would be paid through formation of a voter-approved special assessment district. In November 2012, owners of those 3,855 properties were notified by certified mail that they had until Jan. 25, 2013, to submit any objections to the city clerk. The objections would consitute a "no" vote against the project; lack of response equaled a "yes" vote. Objections from half or 1,928 of the properties would be required to stop the project. The city received 540 objections.[3]

Whether individual property owners chose to actually hook up to the gas was a personal matter. The cost to do so—again, over and above the original assessment per parcel—would be paid by the property owner, the amount depending on each property's distance from the distribution line. Those property owners then faced the additional cost of converting their homes to the new fuel source.

Once the project was completed, assessments by the city were set at $3,265.77. Property owners could pay in a lump sum or in 10 annual payments of $405.27, with an interest rate of 4 percent per annum on the unpaid balance of the assessment.[4] Not paying the assessment by set deadlines would result in foreclosure action. The first payment deadline was Sept. 1, 2015. By that date, 1,089 parcel owners had paid the full assessment; 1,946 made the first annual payment, 171 received low-income deferrals, the assessments for 110 condo units were to be done at a later date, and no payments were received for 607 properties—67 more than opposed the project.[5]

In November 2015, with the conflict of interest exemption lifted, Beth discussed with me her criticisms of the project.

"The need for natural gas is not an equal opportunity for all payers, so the process of requiring residents to contribute to the construction of the distribution system seemed a bit unfair," Beth said. "And there were a number of residents that later indicated they either did not receive their packets or did not realize they had to object in order to oppose the process. That makes it feel like the process was flawed."

In addition, Beth said, fewer residents actually connected to the natural gas distribution system than anticipated, "which seems consistent with the lack of community continuity on the project. Many years down the road, this project will likely be seen as a great decision for the community, but right now it still carries hard feelings for many."

Of other oil and gas activity nearby, but outside the city of Homer, Beth said she had heard concerns from people mostly having to do with water rights and protection of private property. Asked what could be done to help resolve those issues, Beth's comments echoed Barbara's.

"It seems that some type of zoning regulations would prevent the development of wells in otherwise residential areas, but when you get outside of a city structure it is unlikely that any (such) regulations will be developed," she said.

"In the 40 years that I have lived here one thing has been consistent: Whatever is going on that looks like development or progress is going to have a healthy following of those in opposition. There were people that did not like the progress that came with the oil pipeline, they don't like tourism, and they don't like development in Cook Inlet," she said. "It is disheartening to see a way of life change and it usually comes at a price, but when those that are benefiting outweigh those that are not, it is called progress."

When Beth first came to the southern Kenai Peninsula in the 1970s, she said, fishing was the main industry, with the economy just beginning to feel the impacts of oil and gas activities on the North Slope.

"Over the past 40 years, the close attachment of the community and the state to the oilfields has been evident," she said. "The general standard of living is higher, the schools are in better condition and the students have many opportunities that did not exist before oil revenues."

In addition to approximately 90 percent of Alaska's general fund coming from state oil revenues, the oil and gas industry is responsible for one out of every three jobs in the state.[6] When Gov. Bill Walker released his budget for fiscal year July 1, 2015–June 30, 2016, it was clear that having most of the state's eggs in one revenue basket was not healthy. In fact, a drop in oil prices was the reason for the state's $3.6 billion deficit.[7]

"Now we are seeing the downside of being so closely tied to a single-industry economy," Beth said. "As oil prices fall, production goes down and the economy struggles at all levels." However, she added, "Overall, the benefits outweighed the disruption (of nearby oil and gas activities) for me, but there were many that were here before we came that did not feel the same. As generations of Alaskans have an improved life as a result of oil revenues, I think many have changed their minds."

. .

KELLY COOPER was elected to represent Homer on the Kenai Peninsula Borough Assembly in 2014, joining the nine-member board at a time when the borough was keeping an eye on progress of something called the Alaska LNG (Liquefied Natural Gas) Project. As proposed, the project's sponsors—North Slope producers ExxonMobil, ConocoPhillips and BP, as well as pipeline company TransCanada and the state of Alaska—would bring natural gas from the North Slope through an 800-mile-long, 42-inch-diameter pipeline to a natural gas liquefaction facility to be constructed in the central Kenai Peninsula community of Nikiski. Cost of the project was estimated at $45 billion to $65 billion, with the potential to create as many as 15,000 jobs during design and construction and 1,000 for operations.[8] The pipe would be constructed to carry 3 billion to 3.5 billion cubic feet of natural gas per day. The plant would be able to make up to 20 million metric tons a year of LNG, processing 2.5 billion cubic feet of gas daily.[9]

Vincent Lee, TransCanada's manager of major projects, reportedly told the Alaska Legislature's Senate Finance Committee in October 2015 that being involved in the project was "no longer commercially reasonable" for TransCanada.[10] Asked by Sen. Mike Dunleavy, R-Wasilla, if that meant the entire project was not commercially viable, Lee said the project still had potential; however, the company's decision had come after Alaska Gov. Bill Walker indicated an interest in the state having a larger stake in the project.

TransCanada was the force behind the 1,179-mile Keystone XL Pipeline Project that in November 2015 was given a thumbs-down by U.S. President Barack Obama. After pointing out that the project would not make a "meaningful long-term contribution" to the U.S. economy, that it would not result in lower gas prices for Americans, and that "shipping dirtier crude oil into our country would not increase America's energy security," the president said, "if we're going to prevent large parts of this Earth from becoming not only inhospitable but uninhabitable in our lifetimes, we're going to have to keep some fossil fuels in the ground rather than burn them and release more dangerous pollution into the sky."[11]

The same month President Obama rejected the Keystone XL Pipeline, the Alaska Legislature approved a $68 million buy-out of TransCanada's share of the Alaska LNG Project. In December 2015, Gov. Walker made available to the public agreements between the state and two of the pipeline project partners, BP and ConocoPhillips. Walker said the agreements were insurance that there would be gas for the pipeline even if either company pulled out of the project.[12]

Although a decision on the economic viability of the Alaska LNG Project is not expected until 2017, by mid-2015 about half of the 1,000 acres that would eventually be needed had been purchased, according to Facilities Project Manager Michael Britton.[13] Buying up of houses within the facility's proposed footprint also had begun. When I toured the area in November 2015, windows and doors on houses scattered through the affected neighborhoods had been boarded up and "No Trespassing—Ak LNG" signs posted.

Larry Persily, a one-time presidential appointee to the Federal Office for Alaska Gas Line Projects, former associate director of the Alaska Governor's Office in Washington, D.C., and past aide to the Alaska State House Finance Committee co-chairman, was hired by the borough to keep borough Mayor Mike Navarre and the assembly informed about the project.

"The project's impact in the Kenai Peninsula Borough will be more lasting in terms of jobs, in terms of rerouting the highway, its effect on schools and health care. It has a much bigger impact on the Kenai borough than any other jurisdiction on the whole route," Persily has been quoted as saying.[14]

. .

CLOSER TO Homer, having heard complaints from southern peninsula residents who live near the southern Kenai Peninsula's natural gas line but are not in an area where gas has been distributed, Kelly Cooper was instrumental in amending borough code regarding utility special assessment districts. USADs are a process through which residents can finance the extension of main lines of service of public utilities overseen by the Regulatory Commission of Alaska or of city-owned utilities to areas outside city boundaries. Previously, the process required owners of 70 percent of the total number of parcels within the proposed district to support formation of the USAD. Kelly's efforts reduced that to 60 percent "so we could get more USADs passed," she said. The USAD process was used by residents in the Anchor Point area to make natural gas available in areas otherwise bypassed.

On the other hand, Kelly said, she had heard from residents within the natural gas distribution area who are "furious, saying they would never get their money out of it, but people don't understand that it increases the value of their home, makes it more attractive." That's a situation Kelly finds herself in. She has converted her home to gas, but not her seven rental cabins "because of the cost."

"So, this whole topic is right on the fence. You could easily justify the need for it being here for 50 percent of the people. The other 50 percent are saying no, it's not fair, it's forcing government on me," she said of the gas line. "People next to these projects—their property is being hurt and they have no recourse. And there are people being brought into the USADs that can't afford to convert and they have no recourse."

Kelly also considered the overall impact of Outsiders moving to Alaska.

"Those of us that have been here for any length of time, it was because we didn't want to be like the Lower 48, but everyone keeps coming and along with that comes progress and along with that the expectation of a way of life. . . . It's what a majority of people are wanting and that's how our system works," she said.

Requiring a protective barrier around properties might provide a buffer, but Kelly, like Barbara and Beth, recognized that zoning is a four-letter word to many peninsula residents.

"They scream, 'I don't want zoning, don't want anyone telling me what I can or can't do.' That's what they wish for, but when their neighbor does what they want to do, then they're furious. . . . People aren't willing to look at the bigger picture. They only care about themselves and not the bigger world," Kelly said.

As to the benefits of oil and gas, Kelly said, it's all a matter of dollars and cents.

Taxes on oil and gas properties assessed by the Kenai Peninsula Borough "play a vital role in the borough's economy," Mayor Navarre said in his letter to the assembly accompanying the borough's budget for the July 1, 2015–June 30, 2016 fiscal year. The assessed oil and gas properties taxed under Alaska Statute totaled $1.2 billion, Craig Chapman, the borough's finance director, told me by email, with the borough's portion of taxes assessed on those properties totaling approximately $11.5 million. Of that, $5.5 million went to the general fund and $6 million went to the borough's various service areas. Oil and gas revenue in the general fund represents 16 percent of property tax revenue and 7 percent of total revenue in the general fund.

Should the Alaska LNG Project be developed, it is not proposed that it be taxed at assessed value. The three North Slope producers—ExxonMobil, ConocoPhillips, and BP—have negotiated with the state of Alaska to pay $800 million for municipal impact payments during construction and to make $15.7 billion in payments in lieu of taxes, PILT, rather than property tax payments to local governments and the state. That arrangement helped provide the producers with the "elusive fiscal certainty" they were seeking from the state, according to Randy Hoffbeck, commissioner of the Alaska Department of Revenue and chair of the Municipal Advisory Gas Project Review Board. The $15.7 billion PILT is based on a formula that takes into consideration the state's mill rate for the North Slope gas treatment plant and pipeline, the mill rate for the LNG plant and terminals in Nikiski, and a mid-range project cost.[15]

Placing a tax burden on the project, if its capital costs exceed $55 billion, could stress its economics. "The pipeline doesn't have the economics that (the trans-Alaska pipeline) had," Hoffbeck was quoted by the *Alaska Journal of Commerce* as saying.

"The amount the borough will receive if the project proceeds is still in the discussion stage," Chapman said.

Looking beyond the direct impact on the borough's budget, Kelly pointed to the contributions the industry makes to "our little warm fuzzy things," meaning the nonprofits that receive donations.

"The benefit is that they write checks. The number of nonprofits we have, without that we couldn't provide the services we do," she said, and then added, "When you start to actually voice that, it leaves a bad taste in your mouth. You feel like you really have prostituted yourself. And we have. We don't want to admit it, but we have."

While those contributions provide a short-term fix, it isn't worth it in the long term, she said.

"That's the thing I fight so much. When we get these short-term solutions, whew, we don't have to worry about them. But we never look beyond the most immediate emergency. We don't look down the road far enough," Kelly said.

. .

THE AREA BRENT Johnson represents on the Kenai Peninsula Borough Assembly includes Stariski, Ninilchik and Clam Gulch. A commercial fisherman, Brent was born on the peninsula, as was his wife, Judy. After graduating from high school, Brent attended two semesters of college before dropping out and going to work at the Prudhoe Bay base camp of ARCO (Atlantic Richfield Co.) in the mid 1970s. Working as a roustabout, he drove a 20-speed semi box van into the community of Deadhorse, picked up groceries and delivered them back to camp. He also delivered materials around the field and "thought I was hot stuff. Not in too big-headed a way because I knew I didn't know what I was doing, but it was fun," he said. "They loved me. I loved them. The food was fabulous. But . . . I was scheduled to get married to Judy, and I told them, 'Well, I'm getting married, so good-bye.' I'd looked around a little, and the record for young people working away from home wasn't so hot. I thought, looking at the money, how easy it was to make—but the neat thing is I still have my wife."

Trying to find equilibrium between benefits of the oil and gas industry and criticisms he has heard from constituents, Brent said, "There is no code book that says how to balance anything, and I think we have to do that, balance it out. There are sides that would not have oil development at all. That's pretty harsh, especially when you look over the history of Alaska. Oil has certainly been an economic driver big-time."

Brent recommended requiring oil companies to "mitigate their effects on society as much as possible and keep out of real sensitive areas. There's probably places they shouldn't drill."

He also recognized the importance of a diversified state economy and suggested the benefit of reinstituting a state income tax that was repealed in 1980 thanks to the state's accumulating oil wealth.

"I used to have several friends that lived here and worked for the oil companies and now live out of state and just fly up and work at Prudhoe Bay or wherever in oil, so an income tax would catch these people. The same with commercial fishermen who live Outside. Sure, my ideas aren't popular, but lots of times I have unpopular ideas," Brent said.

He is not alone in championing return of a state income tax. In April 2015, Republican Paul Seaton introduced legislation to do just that. Applying to everyone, non-residents as well as residents, who earned money in Alaska, Seaton's tax structure would raise between $600 million and $1 billion annually and repeal a state tax credit on political donations.

"The dire straits of Alaska's budget caused by the precipitous oil price drop requires consideration of diversification of revenue sources. Alaska is the only state without a sales, income or state property tax. The income tax will generate the greatest revenue in an equitable manner at the lowest administrative cost. Although it will not balance the budget alone, individual Alaskan contributions are a necessity for any potential fix, along with additional cuts to expenditures," Rep. Seaton wrote in his sponsor statement for House Bill 182. The bill was introduced April 3, 2015, during the first half of the 29th Alaska State Legislature, and referred to the House Finance Committee. With the convening of the second half of the session in January 2016, the committee's co-chairs, Rep. Mark Neuman, a Republican from Big Lake, and Rep. Steve Thompson, a Republican from Fairbanks, will decide whether to advance Paul's bill.

Paul Seaton in turn is not the only one in state government to see the benefit of reinstating a personal income tax. As noted earlier, Gov. Bill Walker's "New Sustainable Alaska Plan," unveiled in December 2015 and reviewed in his State of the State address in January 2016, included a 6 percent income tax, estimated to equal about 1.5 percent of income for the average Alaskan family and result in projected revenue of $200 million annually.

"A strong, vibrant Alaskan economy requires that we all share in the responsibility to build our future," Walker said in his press release announcing the plan.[16]

. .

MIKE NAVARRE is a second-generation mayor of the Kenai Peninsula Borough. His father, George Navarre, was borough mayor from 1966–1972. Mike served as borough mayor from 1996–1999 and again from 2011 to the present. His office is in the George A. Navarre Building, home of the borough's administrative offices in Soldotna and named for his father. Mike also served in the Alaska House of Representatives from 1984–1996 and was the House Majority Leader in 1989–1990.

Changes driven by the Kenai Peninsula's petroleum industry are a way of life for Mike. He attended junior high school on a split-shift schedule because of all the families moving to the area as a result of booming local oil and gas industry. He's seen dirt roads become paved, streetlights installed and new schools built with oil and gas dollars.

"There have been huge economic impacts in part because of the opportunities that are presented to supply or be a contractor for the oil companies and oil development, but also as a result of just a general economic growth as far as investment for those things," he said. "What's grown up are other businesses supporting goods and services as a result. And then, as people recognize and come to appreciate the beauty and scenery of the Kenai, another industry grows up around that as someone who moved here takes advantage of those things. So it's created a lot of change."

There also have been changes he's observed in the industry itself, some built on past mistakes.

"It is so much better in terms of environmental oversight and regulatory authority because they know it costs them more if they don't address environmental concerns up front," Mike said, recalling a waste dump site in the middle of town when he was growing up, as well as other dump sites that have since been dealt with. "It used to be out of sight, out of mind, and then they found out, wow, there can be ground water contamination. So laws and regulations were put in effect to make sure that you tried to address concerns as much as possible up front."

When I spoke with Mike in 2015, development of the Alaska LNG Project was in the spotlight.

"This project has the potential to be a huge game-changer. There are a lot of aspects to it that are impacts locally, how it's going to affect residents of the Kenai Peninsula and the tax base," he said, listing increased populations within the borough and within schools, development of needed infrastructure and greater emergency response capabilities.

"If you look at the low end, at $45 billion, about 40 percent of that is the plant and dock facility in Nikiski. So almost $22 billion of that $45 billion will be here," Mike said of the $45 billion to $65 billion anticipated cost for the project that begins with a pipeline on the North Slope, stretches across Alaska and ends with the bulk of what's needed—a liquefaction facility, an LNG plant, storage and shipping terminal—in the Kenai Peninsula Borough community of Nikiski.

In November 2014, the U.S. Department of Energy granted Alaska LNG Project the authority to export 20 million metric tons of LNG annually for 30 years to countries covered by free-trade agreements with the United States.[17] In May 2015 the DOE granted the project conditional authorization to export to non-FTA nations.[18]

The Alaska LNG Project has raised questions and concerns from residents near the proposed project. Among them are worries that the project will use up the water supply, that the rerouting of roads will decimate neighborhoods, that eminent domain will be imposed to take over houses and force residents to move.

"I tell them I don't have all the answers and just like them I'm waiting for more information," Mike said. "As we hear from people about specific issues and concerns, we try to follow up on it and find out what's real and what's speculative."

He characterized some complaints as coming from "people who want a particular lifestyle, people who don't want anyone to tell them what they can or can't do, but if someone moves in next door they want to be able to tell them what they can do. And as hard as it is to face up to the fact, what it means is zoning."

Mike recalled when his dad was mayor and tried to introduce responsible planning and zoning. Those efforts were not well received.

Crowley Maritime Corporation,

a petroleum distribution group that delivers freight and diesel fuel to rural Alaska communities, announced on Nov. 10, 2015, that it had been granted approval by the U.S. Department of Energy and the National Energy Board-Canada to import Canadian-sourced liquefied natural gas to Alaska and the Pacific Northwest. The two-year import and export licenses are renewable and allow Crowley to import up to 2.12 billion cubic feet of LNG by truck or by ocean-going vessels.

"It's exciting for many of our commercial and industrial customers because the availability of LNG has been very limited in the past, and it's another opportunity for us to prove that we're a total solutions provider in the energy and logistics industries," Matt Sievert, Crowley's director of business development, LNG, said in a press release posted on the company's website.[19]

The press release went on to say that the company was working to secure long-term, 25-year licenses and was monitoring development of Alaska LNG supply projects that would prove a closer source to the state's interior markets. A day after the press release was posted, the Alaska Dispatch News carried a front-page story about Crowley preparing to begin importing LNG to Alaska. In it, Larry Persily, current advisor to Kenai Peninsula Borough Mayor Mike Navarre and the borough assembly on the Ak LNG Project, noted it was the first time a company had been permitted to import LNG to Alaska.[20]

–McKJ

"So now, as we see growing communities and more development in neighborhoods, now I think a lot of people are waking up to the realization that there's a balance that can maybe be achieved between what government should do reasonably in order to protect some of those areas and lifestyles versus too much government regulation," he said.

Mike was serving in the Legislature when the *Exxon Valdez* spill occurred and the president of Exxon flew in to Juneau to meet with legislators. Considering the possibility of a similar catastrophe in Cook Inlet, he said, "I can't say the *Exxon Valdez* can't happen again, but we couldn't get support to stop all oil and gas because Alaska's economy is so dependent on it and it's so important to the residents and their lifestyle, the opportunity to educate our kids and all kinds of things that have been a benefit. The impacts that we've seen, not all of them are good, but they're part of the overall picture."

Musing about that balance, he told of working in Anchorage after high school graduation and driving back and forth to the Kenai Peninsula on the weekends. The roads were in poorer condition then, and traffic was lighter. Development on the Kenai Peninsula has drawn attention and, as a result, improved highways and increased traffic.

"We have a tremendous amount of people who recognize the Kenai Peninsula for fishing, hiking, sightseeing, boating, all kinds of stuff. And there's more truck traffic because more people live here. ... Sometimes I long for the days when I was a little kid growing up and knew everybody in town and it was a much slower, laid-back lifestyle than we see now."

· ·

AT THE STATE level, Paul Seaton represents the southern Kenai Peninsula in the Alaska House of Representatives. He was first elected to the House in 2002. In the "it's a small world category," Paul is married to my husband Sandy's daughter Tina.

His years in the Legislature have given him opportunities to be involved in the role oil and gas has played, beginning with resolving conflicts associated with shallow gas development and property owners' worries that subsurface rights outweighed surface rights. He also encouraged development of the gas wells on the North Fork Road north of Homer. Paul's commitment to find funding for that multi-phase project overcame two vetoes by Gov. Sean Parnell and led to the city of Homer's project with ENSTAR Natural Gas to provide gas to the area.

Working with the city and the state, he helped Homer secure ownership of subsurface lands around the Deep Water Dock so drilling rigs could tie up. During the time the Endeavour-Spirit of Independence jack-up drilling rig was at the dock, from August 2012 until March 2013, the city realized $577,000 in port and harbor revenues and $181,000 in state oil property taxes.[21]

"I've always tried to make sure anything done is done right and there aren't other impacts that can be mitigated or can be solved," he said. That goes only so far when it comes to one property owner not liking what another property owner is doing, but then again, "sometimes people just don't want to see any change."

I asked Paul if he'd spoken with any of the people I'd met whose lives were so disturbed by noise, lights, vibrations and increased traffic that they felt their reasons for living on the Kenai Peninsula had been destroyed.

"When people want a lifestyle that doesn't want government control and zoning, there's not much you can do," he said. "It's the same way if you're in town and they're going to build a school next to where you're going to be. Whenever you're building something, whether it's a school or a gravel pit or a drilling rig, it's changing the character that some people wanted to maintain. That's what change is."

The plus side of that change, he said, when it comes to oil and gas, is the economic benefits—the jobs and the build-out of local infrastructure. When it came to his efforts to ensure Homer residents had natural gas, Paul weighed the benefits of their having a cheaper, cleaner fuel source available.

· ·

STATE SEN. GARY Stevens, a Republican from Kodiak who represents the Homer area, has been involved in numerous discussions and votes on oil and gas taxes since he served first in the House of Representatives from 2001 to 2003 and in the Senate since 2004: the Petroleum Production Tax Act under Gov. Frank Murkowski; the Alaska Clear and Equitable Share Act and the Alaska Gasline Inducement Act under Gov. Sarah Palin; the More Alaska Production Act under Gov. Sean Parnell.

"Resource development in and around our coastal communities, which comprise the bulk of my Senate district, is an issue of great concern to many constituents," Gary said in response to a question. "The vast majority of these concerns center on potential environmental damage. Also of concern to many constituents is the question of industry taxation, with a common theme being that the resources belong to the state and we should get our fair share from resource extraction."

With no state income tax and with 90 percent of the state's income derived from oil production, "a healthy oil and gas industry is critical to the state," he said. "In large part, it foots the bill for the state's education system, its roads, fisheries, health and social services, and more."

"So, regardless of one's personal feelings about resource development, you can make a successful argument that the industry has benefited the state and local communities," Gary said. Asked if the benefits outweighed any disruptions caused by the industry, he said, "That's really a question every Alaskan should ask themselves. I think you would find the state's population greatly divided on the issue."

. .

DIVIDED CERTAINLY, but Alaska's continued growth and ever-growing dependence on oil and gas lead me to believe "the issue" is heavily weighted on one side. Each of ENSTAR's new natural gas customers on the southern Kenai Peninsula, for instance, approximately 1,500 from Anchor Point south, is another vote in the industry's favor. Another vote for companies to keep exploring. Keep drilling. Keep producing.

Another reason the Paulsruds' home life continues to be disrupted by work on the neighboring drill pad.

One more reason Herff Keith is hearing an increasing hum of activity on the 28-acre pad 1,200 feet from where he built his retirement home.

One more reason Robert and Stacy Correia can be thankful they sold the house they lovingly built and moved their family to a new location.

One more reason there's drilling a half mile from my Ninilchik cabin.

The industry that was the economic backbone of Alaska's bid for statehood now sits in the driver's seat, having secured Alaska's 90 percent dependency. A state that grew by leaps and bounds during boom times and spent liberally to provide for its expanding population now staggers under the financial burden, in desperate need of a fix.

Many of the elected officials I spoke with criticized peninsula residents for their opposition to zoning. Kenai Peninsula Borough rural residents wanting a say in how their neighborhoods develop can pursue local option zoning; borough code outlines the process to petition the borough assembly for consideration as a single-family residential, rural residential, mixed residential, mixed use, industrial mixed use, industrial, or residential conservation district. Any existing uses have to be taken into consideration, however. For instance, according to Max Best, the borough's planning director, if a group of lot owners wanted to create a residential district in an area where oil and gas already had established a presence, the existing oil and gas activities would be grandfathered in.

"You can't go retroactive and make them stop. You can stop expansion, but usually the law holds up that you can't stop them," he said.

The borough incorporated in 1964 and local option zoning became part of borough code in 1966. By then JoAnn Thebaut Steik's father had already signed a lease with Standard Oil. Johnny and Viola Hansen were less than a year away from signing theirs. Oil and gas was well on its way into Kenai Peninsula neighborhoods.

. .

A SCENE FROM THE 2011 action film *Fast Five* comes to mind. In it Brazilian businessman and drug lord Hernan Reyes, played by Joaquim de Almeida, explains the different methods used by the Portuguese and the Spaniards when trying to win control of Brazil from the natives. The Spaniards took a warring approach and lost. The Portuguese arrived with gifts, items Brazil's natives couldn't get on their own and could only continue having if they worked for the Portuguese. Reyes created his own business model after the Portuguese approach.

"I go into the *favelas* and give them something to lose. Electricity, running water, schoolrooms for their kids. And for that taste of a better life, I own them," Reyes says.

What was it Kelly Cooper said?

"You feel like you really have prostituted yourself. And we have. We don't want to admit it, but we have."

BOOM

By Kathleen Bielawski,
©Frozen Music Productions, 1991, www.frozenmusic.com

You like milk and honey, but my cash has appeal,
Business is pleasure — let's make a deal.
This is the land of the free and the home of the brave,
Where a man's a man and a woman's a slave.
It was everyone's party, but I'm nobody's fool,
You hustled your body and you sold your soul.
You wanted it, girl, now you can swallow my pride,
You know you never gave a damn how we do it Outside.

Rape or seduction? Virgin or whore?
I stand accused of corruption, so let's settle the score.
You signed your name on the dotted line;
When the deal was done, you were feelin' fine.
You got yours — and I got mine.
(See.)   (Hear.)   (Speak.)

I spilled the crude, but you spilled the beans.
I got caught with my pants down, but this ain't what it seems.
Your lips said no, but your eyes said yes,
So I ripped off your dress and lay down with you.

Fact or fiction? You say you've been used.
Slight contradiction; let's remember the rules.
You signed your name on the dotted line;
When the deed was done, you were feelin' fine.
You got yours, and I got mine.
(Come!)   (See!)   (Conquer!)

Hard livin' - hard lovin'
We were made for each other: partners in greed.
You got money in the bank and a tiger in your tank,
Now it's my turn, and I'm comin.'

In the land of the free, this case is open and shut.
Here a man's a man and a woman's a slut.
You made your bed, now you can lie where we slept,
But don't dance with the Big Boys if you don't know the steps.

See the Seven Sisters laughing up in the sky,
Watch the camel threadin' through the needle's eye.
Don't cry over spilled milk and stop asking why.
Grow up, baby — you ain't gonna die.

In the game of erection, it's boom or it's bust;
Business is business and I do what I must.
You signed your name on the bottom line;
When the deal was done, you were feelin' fine.
You got yours and I got mine.
See! (No.) Hear (No.)   Speak (No.)

Now the deed is done and I'm feelin' fine;
I got yours and I got mine.
Gonna boogie on down to the Bottom Line.
Come! See!   Conquer!

See (no) Hear (no) Speak (no)
Come! (no) See! (no) Conquer! (no)
Come! (yeah) See! (yeah) Conquer (yeah)
Come! (yeah)   Come! (yeah)
COME!

"On Earth, human activities are changing

the natural greenhouse. Over the last century, the burning of fossil fuels like coal and oil has increased the concentration of atmospheric carbon dioxide. This happens because the coal or oil burning process combines carbon with oxygen in the air to make CO_2. To a lesser extent, the clearing of land for agriculture, industry and other human activities has increased concentrations of greenhouse gases," according to the National Aeronautics and Space Administration.[1]

The long list of consequences includes rising temperatures, increased evaporation and precipitation, melting glaciers and sea ice, rising sea level and warmer oceans. In 2009, Scientific American reported researchers' belief that global warming was responsible for 150,000 deaths each year around the world, that it contributed to five million human illnesses each year, and that the number of related deaths was likely to reach 300,000 by 2030.

Global warming, the article said, speeds the spread of infectious diseases, creates conditions that lead to potentially fatal malnutrition and diarrhea, and increases the frequency and severity of heat waves, floods and other weather-related disasters.[2] 　　　*—McKJ*

LIVING WITH LESS

"The age of fossil fuels will go away,
but they are holding on till the death throes"

PIONEER AVENUE RUNS THROUGH THE CENTER of Homer's small downtown area, its name a tribute to individuals whose pioneering spirit led them to settle on the shores of Kachemak Bay in the late 1800s. Standing just off the edge of the sidewalk, directly in front of Lindianna's Music Garden and Kachemak Bay Wooden Boat Society, is a pioneer of another sort: a renewable-energy-powered street lamp.

"If you want to go green, this is it. You don't have to dig a trench. You don't have to bring a wire to it. Everything is in this one unit," said Kamran Vasseghi, the Alaska dealer for Urban Green Technology, manufacturer of the self-contained unit he is using as a model for what's possible. Kamran's home base is Marina del Rey, Calif. After visiting Homer in 1990 and falling in love with what he saw and the people he met, he began making return visits once, sometimes twice a year, finally purchased a home in 2007, and now spends a week every month in this city of 5,000 near the southern tip of the Kenai Peninsula.

"Our turbines are state of the art," said Kamran, who is also a California distributor for Urban Green Technology (UGT). He ticked off the selling points of the company's four models, beginning with what he described as vertical, rather than horizontal, access, which allows it to operate with wind from any direction. The turbines are quiet. Birds are not attracted to the spinning motion, Kamran said, which results in fewer avian fatalities than other models of wind turbines. The UGT design produces electricity in winds as low as three miles per hour; it is rated to operate in wind speeds up to 120 miles per hour and temperatures as low as 40 degrees below zero Fahrenheit. The turbines are "simple, simple" to install, Kamran said, needing only a team of two people and as little as an hour to put in place. The street light model on Pioneer Avenue combines a wind turbine and solar panels to create enough energy for 365 nights of light, with an additional five days of stored energy in a battery bank. It sells for $10,000, installation runs about $7,500 and freight to Alaska would tack on another $2,000.

"It's totally green. You have zero effect as far as producing any kind of pollution," said Kamran of the design that UGT's website says has "turned wind into the most affordable form of clean energy." He estimates a unit would pay for itself in five to seven years.

His vision for a green future includes the city of Homer installing Urban Green Technology streetlights throughout the city, thereby generating enough power to not only reduce the city's electricity bill but also produce enough excess energy to create a new revenue stream for the city, which could sell the excess power generated to Homer Electric Association, provider of electricity for this part of the Kenai Peninsula. "So with a little investment, the shoe's on the other foot, they're making money," Kamran said.

His vision also includes the city putting a router on every light and providing Internet service to city residents.

"Right now, people at home pay $65-$70 a month for Internet access, but how about if you could have an Internet power supply on each of these and it would be free? You wouldn't have to go to Safeway or the library for WIFI. It would be a win-win situation," he said, adding, "There is so much we could do to grow, so many changes out there. If we really adapt to them, we'd have the perfect spot in this world. And it would be called 'Homer.'"

Is the streetlight getting much attention, I asked Kamran.

"People are interested. They're calling me," he said. "I love this town. That's the reason I bought this and put one at my house, so people can see (them). If I don't sell any of them to anyone here, that's OK. I just feel sorry they don't adapt to it, but they will eventually. It's the future. This might turn their heads around."

· ·

HOMER ELECTRIC Association has been supplying electricity to areas of the Kenai Peninsula, including the southern end, since 1950, when a 75-kilowatt Caterpillar generator met the demand for the cooperative's 56 original members. HEA now has 22,892 member-owners in a 3,166 square mile service area. Eighty megawatts of generation from multiple sources are required to meet electric needs today: Bradley Lake Hydroelectric plant is fueled by water; Soldotna Generation Plant and Bernice Lake Generation Plant are fueled by natural gas; Seldovia Generation Plant is fueled by diesel; and Nikiski Generation Plant is fueled by natural gas and recovered heat, which is HEA's effort to reduce its own energy usage.

"Nikiski is a combined cycle plant, meaning that we are using the waste heat off the natural gas turbine to create steam which is used as fuel to power a steam turbine generator. Using this process, HEA is able to produce up to 18 megawatts of power without using any additional natural gas," Joe Gallagher, HEA's director of member relations, explained. Joe also sits on one of the board committees of the Renewable Energy Alaska Project, formed in 2004 to promote Alaska's use of renewable energy and having more than 80 members representing electric utilities, environmental groups, consumer groups, businesses, Alaska Native organizations, and municipal, state and federal agencies.

Between 2012 and 2015, HEA's annual sales dropped from 489.67 million kilowatt hours of electricity to 465.11 million kilowatt hours, a decrease Joe credits to its conservation-minded members. HEA's annual Energy and Conservation Fair is a platform for vendors and energy conservation experts to share energy reduction opportunities with members. The HEA website also offers energy-saving tips for home and business owners. A student contest associated with the fair provides an energy focus for youngsters, as do visits by HEA's engineers to junior high and high school science and technology classes to discuss energy generation, renewable energy and efficiency.

"Our members are much more aware of energy conservation, and there is also an improvement in the efficiency of appliances, lighting and other electric products," Joe said, adding that a trend of warmer Alaska winters also has had a role in how much energy HEA sells.

Under the heading of renewables in HEA's board policy is an impressive list of possibilities: hydroelectric projects, conservation efforts, and recovered heat, excluding generation facilities; net metering, an avenue for members to reduce the amount of electricity they purchase from HEA by interconnecting member-owned or leased generation facilities; and wind, solar, geothermal, waste-to-energy and other recognized methods. HEA policy recognizes "the importance of the environment and the need to provide power for sustainable growth" and is "committed to the continued development of renewable energy sources." Thanks to the Bradley Lake Hydro Project the co-operative's renewable energy has exceeded

15 percent of HEA's annual peak demand. The goal is to increase that to 22 percent by the end of 2018.

According to Joe, HEA's primary focus on renewable energy at present is the proposed Grant Lake Hydroelectric Project, which would provide five megawatts of energy. After five years of study, HEA planned to submit a final license application to the Federal Energy Regulatory Commission in 2015. However, preparing the application took longer than anticipated, Joe said, and the plan now is to submit it during 2016. If it's approved, the co-op's next step will be to secure an estimated $58 million for construction. Mike Salzetti, manager of HEA's fuel supply and renewable energy program, has said that funding from the Alaska Energy Authority for the project would be welcomed. If construction costs were to be paid by HEA, they would be financed over the length of the project's federal permit so as not to increase rates to HEA members.[3]

The renewable-energy grants program of the Alaska Energy Authority allotted more than $2 million to Kenai Hydro, a subsidiary of HEA, to investigate four potential hydro projects near Moose Pass, a community on the eastern side of the Kenai Peninsula. Grant Lake is the only project of the four still being considered. The energy authority denied construction funding for Grant Lake in 2011, citing among other reasons public opposition to the project.[4]

"Hydro has the worst ongoing impact (on salmon) of any of the renewables," said Moose Pass resident Mike Cooney at a November 2014 public meeting on the project. Cooney asked HEA representatives if they could give a number for how many salmon return to Grant Lake to spawn. According to *Seward City News* coverage of the meeting, none of the fish biologists and engineers on the eight-person HEA team of representatives at the meeting could respond to Cooney's concerns.[5]

Seward resident Mark Luttrell criticized the plan for a road to the project that would "bring all the bad habits of humans just a little bit deeper into the wildland" and cautioned about vandalism to historic artifacts and the impact of "an industrial facility there in that small town" of Moose Pass.[6]

. .

HEA IS EXPLORING tidal power as well, through its partnership with Ocean Renewable Power Co.

The technology is still very much in the research and development stage, but ORPC has installed the country's first grid-connected tidal energy generator in Maine, and they are studying the feasibility of a tidal project in Cook Inlet," Joe Gallagher said.

Working with the Alaska Energy Authority, HEA is also supporting a proposed Battle Creek Diversion into Bradley Lake, which would result in about a 10 percent increase in the amount of renewable energy delivered from that facility.

. .

OTHER ALASKA communities are looking as well to alternative solutions for their energy challenges. Kodiak, the Kenai Peninsula's island neighbor to the southwest, has made headlines with its reliance on wind and water to provide power. Kodiak Electric Association's main power source, 81.5 percent, is three hydroelectric turbine generators; the Pillar Mountain Wind Project provides more than 18.3 percent, and the remaining 2 percent is provided by diesel generation.[7] On the tiny Bering Sea island of St. Paul, Tanadgusix Power (TDX), a subsidiary of the village's Tanadgusix Corporation, has announced plans to have St. Paul fueled by 80 percent renewable energy by 2020. It developed a hybrid wind-diesel power plant to reduce local energy costs in 1999, has installed six 225-kilowatt wind turbines, and plans to expand the system to meet its 2020 goal.[8]

BILL STEYER OF Homer is one of more than 40 independent contractors statewide certified to evaluate home energy performance and listed as Alaska Housing Finance Corporation-approved energy raters. Bill's audits include an assessment of a home or business's energy performance, recommendations for increasing energy efficiency, and "basically trying to help businesses and people save on their bottom line by cutting their energy overhead costs," Bill said.

Interest in having an audit done has dropped in the past two years as natural gas has been made available to Homer area residents and resulted in lower energy costs. A drop in oil prices also has caused a decrease in calls asking for Bill's expertise. The decline is something he anticipated.

"If you do a renovation to your house (to save energy), maybe put something in the crawl space, you might be able to save $120, maybe $200 a year by putting in some foam board that might cost $200 to install," Bill said. That same repair might result in only a $60 savings for someone already using a less costly energy source, such as natural gas, "so there's less of an incentive for people to want to do an audit."

While he understands that reasoning, Bill said, he also knows that decisions based on the pocketbook don't do anything to reduce the carbon footprint created by the use of fossil fuels.

As an energy rater, Bill's first step toward finding energy savings is looking at opportunities for a quick fix and then at renewable-type energy sources. He used the example of one Homer business that realized a $6,000–$7,000 annual savings by simply switching to more energy-efficient lamps.

"The cost of doing that was about $7,000. In an energy-efficient world, if you can get three years or less in an energy payback, it's usually economical and usually an incentive," Bill said. He compared that to the cost of installing photovoltaic panels to convert solar energy into electricity. "In Alaska they might have to pay

$40,000 or $80,000 for panels that will only help out with 10–15 percent of electricity costs because the sun is not always at the same angle, plus we don't have as much solar potential. So, what's more cost-effective? That's how it typically works. You have to look at every situation individually. I'm an advocate of renewable energy, but for every assessment I might do I want to give them the full menu of options from the most cost-effective to least cost-effective to help someone make the decision."

In his own home, Bill installed an $8,000 solar-thermal heating system "because I believe environmentally, for the planet, it's a good thing to do to reduce the carbon footprint. But not everybody has the ability from the resource standpoint," he said.

His experience as a Peace Corps volunteer in Nepal and in other travels around the world have led him to advocate for behavioral changes that cost nothing but lead to savings—something as simple as turning lights off when you leave a room. Using Nepal as an example, he said, "In a room one light is enough. They don't have the culture of waste that we have because we've had so much energy. We overkill with lighting."

On the other hand, from his work in remote Alaska settings Bill also is well aware of the unique challenges the state poses, which result in higher energy costs that in turn result in an even greater need for more energy efficiency.

· ·

WHEN HOMER photographer, author and outdoor adventurer Taz Tally's home energy costs began to skyrocket in 2008, he sought help from the Alaska Housing Finance Corporation's energy audit program.

"I implemented all the things the audit recommended and my energy bill dropped 40 percent. That is not insignificant," said Taz.

The program is limited to homeowners, however, and Taz was inspired to seek out help for energy-aware business owners.

"I started looking around, asking questions, meeting people and found the U.S. Department of Agriculture had energy programs for small businesses," said Taz, who was awarded a $100,000 USDA grant to focus on small business energy needs. "We provided audits for 27 businesses in Homer, and there was an average energy savings of about 40 percent. The USDA considered it to be a success. They got participation out of it, and people actually did something with the information."

That success led to Taz being awarded a second grant with the focus extending beyond Homer to the entire Kenai Peninsula.

"We did 23 audits for small businesses and it was a similar success. Businesses anted up and invested in energy efficiency," Taz said. "We were able to document the dollars spent (on improvements) because we provided the audits and then helped small businesses write their own grant requests to USDA to help pay for the recommended improvements, at 25 percent of what they would invest."

The rapport Taz was building with USDA led to a third opportunity.

"They'd been wanting to go to Native villages for years, but that was a whole different ball of wax, a different culture, different everything, so they encouraged us to write a grant and we got $200,000 to go to Nome and Bering Strait communities and we did 40-something audits for small businesses out there," he said. Again, the effort was successful "in that there were people who didn't think we'd get the villages to participate at all, but we were able to do that."

Continuing to build on that momentum, Taz then received two $200,000 grants from Alaska Village Electric Cooperative, electricity provider for 51 locations throughout rural Alaska. When he and I met during the summer of 2015, he and his crew of three were in the process of shepherding 80 small businesses through the process, "trying to get them to invest in their audit-recommended improvements."

Echoing Bill, Taz said "the lowest hanging buck" is the reason for doing energy efficiency audits. "It's been proven time and time again that first you employ energy efficiency and then do renewable energy. You want to reduce total overall energy, and that reduces what you need for renewable energy."

Renewables are relatively expensive, he said, "depending on how big a view you take. ... When you and I make decisions about where to use oil or gas or put in a wind turbine or solar energy, we look at the direct cost in the short term. But in terms of society as a whole big picture, if you consider probably trillions of dollars global warming is going to cost, you can make the argument that actually renewables are less expensive and make more sense."

From his involvement in energy reduction, Taz believes oil and gas development "is going to be short- to mid-term because ultimately it's too expensive." In its place will be distributed energy, "where every house or building will start to generate more and more of their own energy, whether electricity or wind or solar." Although he sees himself as "a very small player in a very big state," he also knows the difference one person can make.

"If everybody does their part, if everybody does a little bit, we'll accomplish enormous things," Taz said.

· ·

WORKING OUT of the Kenai office of Wisdom and Associates, Inc., Rob Moss performs inspections for real estate transactions, for the buyers as well as the sellers. He also does inspections for code compliance for banks. Since 2002, he has conducted energy audits, both residential and commercial, across the state.

"I actually find it pretty interesting, particularly for commercial because there are so many different challenges and areas to look at for improvements and to be able to help people out," he said. As with Bill and Taz, Rob sees his biggest goal on the consumer level as reducing energy consumption and saving money.

"But on the larger scale, one of those big goals is to reduce overall consumption in the state."

Also like Bill and Taz, Rob said cost is the driving factor when it comes to a client wanting an audit done.

"Certainly there are people interested in going green for the sake of it, but economics still is the main driver," he said. "Green technologies haven't hit critical mass here. The cost isn't quite yet low enough for them to be widely implemented. I think their day is coming for sure, but right now there is still a huge outlay."

From his travels around the state, Rob sees communities attempting to implement programs less dependent on fossil fuels, "but really they're still running into a challenge with quality of the power grid and being able to maintain those systems in harsh environments," he said, adding that what will eventually bring green technology to the forefront is a "serious economic crunch, where dollars and cents make it economically feasible to make that switch. I think that will be the main driver. Secondarily, people are just becoming more concerned about it. . . . We're getting there, getting closer every day. A lot of people like going green or alternative, but it definitely comes down to the pocketbook."

· ·

SCOTT WATERMAN manages the energy programs in the rural research and development division of Alaska Housing Finance Corporation (AHFC), a self-supporting public corporation. Since 1986, AHFC's mission has grown beyond providing affordable loans and now includes public housing programs, energy efficiency and weatherization programs, senior housing programs and professional development. According to its 2014 annual report, AHFC recognized reduced energy as "the No. 1 way to save Alaskans money. AHFC weatherized 3,360 homes making them safer and more energy efficient, with the average annual savings totaling $1,300 per year per family."

AHFC sponsors a number of energy programs: Energy efficiency rate reduction, which offers reduced interest rates when financing new or existing energy-efficient homes or when borrowers buy an existing home and make energy improvements. An energy efficiency revolving loan fund for public facilities, which provides financing for permanent energy-efficient improvements. Rebates of as much as $10,000 in home energy improvements, and rebates up to $10,000 for Alaskans building or purchasing newly constructed, energy-efficient homes, plus second mortgages for energy improvements on owner-occupied properties and weatherization assistance for Alaskans within specific income limits.

Scott has been with AHFC for 25 years. He oversees all the corporation's energy efficiency programs except for weatherization, home energy rebates, and a supplemental housing assistance program. AHFC also offers classes on energy efficiency.

"Combatting energy illiteracy is one of our primary focuses," said Scott. "We try to offer the ability to understand energy efficiency in buildings to basically anybody that uses a building in any way—from homeowners to people that maintain or operate a building, investors that are financing a building, builders, people that are repairing or retrofitting a building, architects, engineers, plumbers, electricians. All of those different people can have a significant role in the energy use and misuse in a building."

What difference does it make to use energy more wisely?

"First and foremost is cost. If your weekly energy bill is a hundred units, dollars, gallons, whatever, and you can get by with 85 a week, or 75 or 70, which would you choose? There's really not any reason to spend 100 units every week if you have a choice. A lot of people don't realize they have a choice, so the first part is just educating them that they do have a choice and then how to go about exercising that choice," he said.

While cost is a primary motivator for Alaskans wanting to reduce their energy consumption, Scott listed comfort as another factor, as well as health and safety, and durability, which in Alaska includes a structure's suitability under harsh weather conditions. Of those factors, Scott said, comfort comes first.

"They'll say they started (the energy-reduction process) for the money, but what they didn't realize is how much more comfortable their home can become," he said, using the example of one homeowner who was having her home draft-proofed. "The first day she came home and found it was a little warmer and she hadn't touched the thermostat. The second night it was even warmer. The third night it was very comfortable. She had no idea that would be the benefit of sealing up drafts."

Scott looks at Alaska as several energy states within a state, each with its own pluses and minuses. Southeast Alaska, known as the panhandle, has plentiful and cheap hydroelectric power but high oil costs. Fairbanks, in the heart of the state, has high fuel oil and electric rates. Costs for fuel oil and electricity in remote Alaskan communities have increased exponentially. The Kenai Peninsula falls within an area that has relatively affordable electricity and cheap gas, "so heating is not really that much of a factor," he said. Until recently, he said, with the introduction of natural gas, Homer "had both the most expensive fuel oil and moderately high electric rates."

Finding and sealing drafts was most often called for in the weatherization and home energy rebate programs, Scott said, followed by insulation and updated appliances.

He didn't have a gauge for how energy consumption in Alaska overall has decreased as a result of AHFC's programs, but he could point to the net effect the weatherization and rebate programs have had on the 42,000 homeowners who have benefited from the programs since 2008.

"That number is $60 million a year returned to the economy in direct savings," he said. "That means if you write a $70 check to Homer Electric Association instead of $100, that remaining $30 you're much more likely to spend in Homer and benefit the economy."

In order to achieve that $60 million in savings, AHFC spent about $515 million, which equates to about an eight-and-a-half-year buyback.

Scott estimates that 64 percent of homeowners actually follow through with recommendations resulting from an audit. "We saw a big demand for the programs when the price of oil shot way up in 2008 and the programs were first announced, but we did not see a huge falling off this last year when oil prices fell. There's an average of 500 requests a month statewide."

As to how Alaska compares with other states, the 42,000 homes audited since 2008 represent about 16 percent of the total occupied homes in the state, Scott said. Even considering the percentage of people who follow through on recommendations, "I doubt any other state has done as high a percentage of home retrofits."

On the other hand, the nonprofit American Council for an Energy-Efficient Economy scores Alaska 43rd out of 50, based on the participation of its utility companies.

"In Alaska, our utilities are not interested in energy efficiency in any meaningful way with a slight exception for Golden Valley Electric Association in Fairbanks," said Scott. "There's a process called 'de-coupling' that other states have employed where the incentive is not just to sell more electricity or gas (but instead) to deliver their product efficiently. In Alaska, the only incentive is to sell more."

He drew a direct link between energy efficiency and alternative and renewable energy.

"The cheapest BTU or kilowatt is the one you don't use. So if you're looking at designing an energy system, whether for a house, business or whatever, the less you use, the smaller the system you have to use to deliver it. So that's where energy efficiency fits in," he said. "Efficiency is very important when it comes to renewables."

Scott's 2,500-square-foot house on Diamond Ridge, above Homer, is a good example. Although he works in Anchorage, he spends most weekends at his home on the ridge.

"We just got our propane bill. That's the only fossil fuel we use in that house other than electricity. It was $79 for two years and part of that was the tank rental," he said. "What we did was put in solar thermal that provides virtually all our heat and hot water from mid-February until near the end of October. Then we have a wood-fired boiler that takes up the slack from November to January and we usually use three to four cords of wood a year."

For perspective, he compared it to a neighbor's home where the owner spends between $2,000 and $4,000 a year for heat.

"I'm not a zero net energy house, but I'm probably closer than many people are," Scott said.

· ·

THROUGH HIS firm Deerstone Consulting, Brian Hirsch offers expertise in energy resource assessment, planning and performance. In 2008, Deerstone worked on the city of Homer's climate action plan to help reduce the city's energy and fuel usage. In 2009, Brian became Alaska's representative for the National Renewable Energy Laboratory (NREL) based in Golden, Colo.

NREL was established in 1974 as the Solar Energy Research Institute. In 1991, under President George H.W. Bush, it was designated a U.S. Department of Energy national laboratory and its name changed to National Renewable Energy Laboratory.[9] Managed under contract with the DOE by the Alliance for Sustainable Energy, NREL's mission is to "develop clean energy and energy efficiency technologies and practices, advance related science and engineering, and provide knowledge and innovations to integrate energy systems at all scales."

One example of NREL's work began on May 4, 2007, when a deadly tornado touched down in Greensburg, Kansas. Rated an EF5, the tornado was 1.7 miles wide, stayed on the ground for almost 29 minutes and had recorded surface winds as strong as 205 miles per hour. It killed 11 people, injured 63, and destroyed the town.[10]

In piecing Greensburg back together, local leaders chose to make it a model of sustainability, with the help of the Department of Energy and NREL. This resulted in more than half the new homes being rated as using 40 percent less energy than code on average and gave Greensburg the highest per-capita concentration of LEED (Leadership in Energy and Environmental Design) certified buildings in the United States, with 13 showcase buildings saving a combined total of $200,000 in energy costs annually, plus a 12.5 megawatt wind farm to produce enough energy at the time to power every house, business and municipal building.

"They decided to rebuild their community as clean as possible. NREL was the front-line implementer," said Brian.

• •

As NREL's SOLE staff member in Alaska, Brian worked all over the state. One of his involvements was with Cook Inlet Region Inc.'s Fire Island Wind project. Located about three miles offshore from Anchorage, it is the first wind-powered supplier of electricity to the city and provides power to 6,500 homes. Consisting of 11 wind turbines with a capacity of 17.6 megawatts, the project was completed in 2012 and has eliminated the need for an estimated 500 million cubic feet of natural gas each year.

"We did the analysis that kind of made everyone comfortable with making the investment," said Brian.

His work with NREL also took him to several other countries, including Indonesia, where he was surprised to find some simi-

281 • Living with Less

Crude oil is the most commonly traded

commodity, according to InvestorGuide, along with its derivatives including heating oil and gasoline; natural gas holds ninth place.[11] Banks and diversified financials dominate Forbes' Global 2000, a list of the world's largest, most powerful public companies, using revenues, profits, assets and market value as a yardstick. Second place goes to oil and gas, with 136 companies making the cut.[12] ExxonMobil and Chevron are among Fortune 500's top 10 companies with the highest profits.[13] An online source, oilprice.com, listed the following revenues for oil and gas companies in 2014: Saudi Aramco, $378 billion; Royal Dutch Shell, $419.4 billion; ExxonMobil, $364.8 billion; Petrochina, $367.94 billion; BP, $353.57 billion.[14] With the price of oil dropping below $50 a barrel, the International Energy Agency reported the demand for oil had increased at its fastest pace in five years, by 1.6 million barrels a day during 2015, predicted to continue at a rate of 1.4 million barrels a day in 2016.[15]

Another way to measure the petroleum industry's influence is by looking at its presence in everyday life. Fuel for automobiles and lights for dark rooms are just the surface when it comes to plumbing the depths of humanity's oil and gas addiction. Consider this short list, prepared by the U.S. Department of Energy's Office of Fossil Energy, of items that come from oil and natural gas:

Air mattresses, ammonia, antifreeze, antihistamines, antiseptics, artificial turf, artificial limbs, aspirin, awnings, balloons, ballpoint pens, bandages, beach umbrellas, boats, cameras, candles, candies and gum, car battery cases, bar enamel, cassettes, caulking, CDs/computer disks, cellular phones, clothes

→

line, coffee makers, cold cream, combs, computer keyboards, computer monitors, cortisone, crayons, credit cards, curtains, dashboards, denture adhesives, dentures, deodorant, detergent, dice, dishwashing fluid, drinking cups, dyes, electrical blankets, electrical tape, enamel, epoxy paint, eyeglasses, fan belts, faucet washers, fertilizers, fishing boots, fishing lures, fishing rods, floor wax, food preservatives, footballs, glue, glycerine, golf bags, guitar strings, hair curlers, hair coloring, hand lotion, hearing aids, heart valves, house paint, hula hoops, ice buckets, ice chests, ice cube trays, ink, insect repellent, insecticides, life jackets, lipstick, loudspeakers, luggage, model cars, mops, motorcycle helmets, movie film, nail polish, noise insulation, nylon rope, oil filters, paint brushes, paint rollers, pajamas, panty hose, parachutes, perfumes, permanent-press clothes, petroleum jelly, pharmaceuticals, pillow filling, plastics, plastic toys, plywood adhesive, propane, putty, purses, refrigerants, refrigerator linings, roller skate wheels, roofing, rubber cement, rubbing alcohol, safety glasses, shag rugs, shampoo, shaving cream, shoe polish, shoes/sandals, shower curtains, skateboards, skis, soap dishes, soft contact lenses, solvents, sports car bodies, sunglasses, surf boards, swimming pools, synthetic rubber, tape recorders, telephones, tennis rackets, tents, tires, toilet seats, tool boxes, tool racks, toothbrushes, toothpaste, transparent tape, trash bags, TV cabinets, umbrellas, unbreakable dishes, upholstery, vaporizers, vinyl flooring, vitamin capsules, yarn.[16]

larities with Alaska, considering that one area is tropical and the other arctic.

Like Alaska, Indonesia has hard-to-reach locations, he said. "They have remote diesel generators, fly-in communities, driving on little islands. So the technologies are very similar, and renewable energy efforts are very similar."

Before his time with NREL, Brian spent more than 25 years working with tribal and community governments, corporations and universities on issues of renewable energy and energy efficiency. He served on Homer Electric Association's board of directors. He holds a Ph.D. in natural resources from the University of Wisconsin-Madison, with a focus on energy analysis and policy.

The projects in which he's been involved and the places he's traveled, throughout Alaska, Canada, Central America and Indonesia, have taught Brian that being weaned off fossil fuels isn't a question of having the right technology. Nor is the cost of switching to renewable energy—energy that is generated from natural resources such as sunlight, wind, rain, tides or thermal heat—the problem.

"It's really kind of a people problem more than a machine problem," he said. "I remember as a young adult being intrigued by the technology side of things. There was so much we didn't understand and (the conviction was) if we could figure it out it would be easy to do the rest. Sure there were challenges, like changing people's minds, but presented with good information, that would take care of itself. But as I've seen these many years now, that's not the way it is at all. It's the exact opposite. It's much harder to change people's minds."

That might be true in other areas as well, he said, but there are several factors that set energy apart.

"There's the scale of energy in general. Energy commodities are the largest industry in the world," he said. And in addition to the industry's sheer size, Brian pointed to the key role oil has played and continues to play worldwide.

"You look at the evolution of us as a species, especially over the last 150-plus years. The industrial revolution was created by mining coal out of the ground for energy. . . . To mine coal out of the ground they basically created pumps and engines and water displacement and began to master thermal dynamics, and that spurred new forms of energy and new ways to ravage our planet and harvest more of it," he said.

Now offshore drilling drives the "absolute most sophisticated technology," Brian said, noting what it takes to precisely position an offshore drilling rig so it can access the treasure buried some 20,000 feet beneath the sea bed. "The fact that renewables have even made the kind of inroads they have is fairly astounding."

"There are many people on the front lines right now that are not going to win. That's the ugly reality," said Brian. "At the same time, there certainly are cases of individual situations where oil companies have lost. Bread and butter grassroots organizing. Fighting in the policy realms. Community meetings. Legal and technical challenges. Anybody who has something like that going on is up against a huge, massive monster that wins a lot, is arrogant, greedy and has really high demands, (but) if you slow them down or cost them a lot of money or are a pain in their side, then they may just go somewhere else. That doesn't stop the thing, but in terms of applying pressure to the wound, that's the first place you start if you're talking about a well in your backyard or front yard or next door neighbor's yard."

With low-cost, abundant, reliable renewable energy as the long-term goal, he said, there are multiple ways to get there. One strategy is "to essentially put a choke on the transportation of these fuels such that their prices go up. We are absolutely right now in this really interesting transition where renewables are starting to become as inexpensive as or less expensive than fossil fuel. (But) the rule of thumb is that one gallon of diesel fuel provides 2,000 hours of labor. That's one person's work for a year. That's in one gallon of fuel and that's mind-blowing. That's what we're up against, but we're getting there. And as the transition happens, we won't look back. The age of fossil fuels will go away, but they are holding on till the death throes."

"We'll probably poison ourselves and burn up this planet before renewables win," Brian said, causing me to look up from my note-taking to see if he was overstating to make a point and perhaps smiling. He wasn't. "But if we can survive ourselves, the long-term trajectory is that renewables will beat these dark forces. If we can manage it, that's where we'll go—and you have to fight every day on it. Some people are warriors in the trenches. Some people are big thinkers. Some manage incremental change. None of them are better than others. They're all necessary and essential for—hopefully—getting us out of this mess."

· ·

NEWS ON DEADLINE: The Alaska SeaLife Center in Seward has eliminated 98 percent of its fossil-fuel-dependent heating needs. The new source of heat: ocean water. The center announced in April 2016 that it had successfully installed CO_2 refrigerant heat pumps to replace oil or electrical boilers, for an estimated savings of as much as $15,000 a month and a 1.24-million-pound annual reduction in carbon emissions.[17]

The SeaLife Center is a marine research and wildlife rehabilitation facility that opened in 1998 with partial financing from *Exxon Valdez* Oil Spill (EVOS) criminal settlement funds.

Bless Me, Hills

A late fall storm last night brought the season's first snowfall. Confirming what the chill creeping into the cabin had suggested, the accumulating white groundcover glowed in the darkness when I checked before going to bed to make sure the cabin door was locked. Warm air from the small stove on the main level was being pushed by the gathering coolness up to the loft. Knowing it awaited and that I'd soon be under layers of blankets, I turned down the heat and climbed the stairs.

This morning the clouds had cleared and a soft blue sky spread overhead. Temperatures had dropped into the 20s. Last night's wet snow now sparkled beneath a sun rising toward its mid-day zenith. As on most mornings here at the cabin, my thoughts turned to the beach. Standing on the porch, I could hear the surf rolling up the shore of Cook Inlet, inviting me to draw near.

The curved mile or two of beach that separates the mouths of Ninilchik River and Deep Creek was rimmed with crunchy snow and ice. From the angle of the waves, it was evident that the tide was on its way out. The slowly receding water offered bare gravel for less treacherous walking and uncovered shells, agates and bits of beach glass otherwise hidden by the snow.

The mountains on the far side of the inlet were brilliantly white in the morning sun. The waves rolling in and out were still gray from yesterday's storm. A chilly breeze from the north convinced me to pull my sweatshirt hood over my knitted cap for more warmth. Someone in a pickup truck was parked on the other side of Deep Creek, but on this side I had the beach to myself.

By the time I returned to the cabin, the near-noon sun was pouring through the trees, dancing on the thin white groundcover. More snow will come, but today the summer grasses still stand tall. Dried stalks caught the morning light, a golden dividing line between the white snow and blue sky.

It was on a morning much like this that I happened upon a prayer by Native American poet Luci Tapahonso and tacked it to the wall beside the windows:

Bless me hills
This clear golden morning
For I am passing through again. [1]

My Roots IV · Writing

THE DECISION TO CUT MY FINANCIAL TIES to Alaska's oil patch and move toward a quiet life of writing from the solitude of the cabin felt much less momentous next to the shock of Mom's unexpected death.

I was finally at the cabin fulltime, but the thrill of that freedom and the future I'd eagerly imagined were overshadowed by the absence of Mom's kitchen light shining through the trees in the evening, her phone call inviting me to stop by for a game of Scrabble, or the comfort of walking the short distance from my front door to hers for a cup of mocha, her beverage of choice to start the day.

Out of the emptiness, a new outline for my life took form as I immersed myself in a recently formed nonprofit, Ninilchik Native Descendants. Memories shared by Dad and other villagers of his generation illuminated a slice of history many of my generation had never known. Details of who arrived when and from where, who married whom, where they lived, began fitting together. The information I was collecting about Dad's side of our family soon filled enough pages to paper the cabin's bathroom door with charts.

When my visiting 4-year-old grandson, Colby, Emily and Joe's son, realized these were the names of individuals on only one branch of his family tree, his hazel eyes opened wide with amazement. Connections to other village families made evident the tight warp and weft of Ninilchik's history, my history, Colby's history.

The cabin began filling with books about Ninilchik and the Kenai Peninsula. Pictures and maps made their way onto the walls. Jars, vases and baskets overflowed with beachcombing treasures. Curiously shaped pieces of driftwood sat drying on the deck. The cabin proved itself more than a uniquely shaped roof over my head or a reassertion of my claim to a piece of the family homestead. It was home in the fullest sense of the word.

That fall a gathering of elders and young people was held in Kodiak as part of a collaborative project between the Smithsonian Institution's Arctic Studies Center and the Alutiiq Museum and Archaeological Repository, located in Kodiak. Ninilchik Native Descendants was invited to participate, and elder Maggie Rucker and 15-year-old Tiffany Stonecipher were selected as our representatives. I had the honor of accompanying them and giving a presentation on the village history I was learning. This project stretched across several years and resulted in "Looking Both Ways: Heritage and Identity of the Alutiiq People," a museum exhibit and an accompanying catalog for which I was invited to write several pieces.

This was gratifying from more than one perspective. But the goal of living as a writer required careful planning. As a wet-behind-the-ears freelancer, it was clear I needed to know something about operating a business. Enrolling in a small business management certification program offered through the Kenai River campus of the University of Alaska Anchorage proved a good step. A course in writing business plans helped define my writing goals. Classes in bookkeeping gave insights for tracking revenue and expenditures. A marketing course led to an article about bed and breakfast inns on the Kenai Peninsula that I sold to *Alaska Business Monthly*.

That first sale inspired me to look for more story ideas and other publications. In addition to articles appropriate for *Alaska Business Monthly*, I wrote feature stories, travel pieces and articles for seniors. I learned the writer's trick of making the most of my time by turning one interview into multiple stories that could be sold to more than one publication. For instance, an interview with Jean Keene, known before her death in 2009 as the "eagle lady" of Homer, Alaska, provided material for three articles: A feature piece for the *Peninsula Clarion*, a daily newspaper out of Kenai, focused on Keene's use of scraps from a fish processor to feed from her front yard the hundreds of bald eagles wintering on the Homer Spit and on the flood of photographers it drew to Homer, and some about her colorful background as a trick rodeo rider. For the monthly *Senior Voice* I stressed the strength and tenacity it took for Keene, who was in her 80s, to feed the eagles in spite of bitterly cold weather, as well as the system she'd developed for finding and storing the frozen fish scraps, and the crew of volunteers who helped her. Keene's struggle with breast cancer became the theme of a piece for *MAMM*, a national breast-cancer survivor magazine published in New York.

With each byline I gained confidence and began to push the horizon of story possibilities. When I heard that my cousin's husband, Steve Okkonen, was part of a scientific team aboard a U.S. Navy nuclear submarine sailing from Pearl Harbor to the Arctic Ocean to study what was happening in that little-understood body of water lying beneath a cover of ice, I wrote to the "for more information" email address listed on the website and asked to be included in one of three groups of journalists the Navy was inviting to witness the submarine's surfacing at a research station located on ice covering the Arctic Ocean north of Barrow.

"Who are you writing for?" was the first question asked by Lt. Steve Mavica, the Navy media person.

Desperate, I threw out the name of George Bryson, "We Alaskans" editor for the *Anchorage Daily News*, who had accepted my 1992 story about Dad's and my trip to Russia. I hoped against hope that George would take this story if the Navy gave me its approval.

"Are you bringing a shooter?" was Mavica's next question.

Realizing he must mean a photographer, I gave the best non-answer I could think of.

"That hasn't been decided yet," I said.

It worked. In April, I flew with *Anchorage Daily News* photographer Erik Hill and other U.S. and foreign journalists to Barrow, where we were briefed by the Navy. The next morning, dressed in cold-weather gear, with pens replaced by pencils (our pens, we were told during the briefing, would freeze in these sub-zero temperatures) and with camera gear wrapped in layers of insulation, we boarded an Era Aviation Twin Otter for the 150-mile flight to the Applied Physics Laboratory Ice Station (APLIS) climate research station.

After landing on a tiny available space of six-foot-thick ice between two pressure ridges, we received another briefing by the station's staff. We could interview and photograph anyone we chose but were warned to stay close to camp to avoid encounters with polar bears. That night we bunked in prefabricated plywood structures erected for the scientists and camp crew. At 4 a.m., we scrambled out of our sleeping bags and stood by as Commander Robert H. Perry and his 120-man crew raised the 292-foot vessel through the frigid depths and the ocean's icy covering.

My article and Hill's beautiful photographs were paired in "The Vanishing Ice: Scientists, submarine probe polar warming," the cover feature of the May 9, 1999, "We Alaskans."

In November, after several of my stories had appeared in the *Peninsula Clarion*, Lori Evans, the executive editor, told me there was an opening on the staff and asked if I would be interested in working in a newsroom.

"I've never considered being a reporter," I said, surprised by her offer. "I'm not sure I can do this."

"We're not sure you can either, but let's give it a try," was Lori's straightforward response.

Working on a daily basis with an editor was a dream come true. It offered regular input to improve my writing. The *Clarion*'s peninsula-wide distribution was an avenue for me to become recognized as a writer. I also was constantly developing story ideas as a reporter on which I could expand as a freelancer.

The financial benefits of journalism didn't compare to what I'd made in the oil industry, but the honor of hearing and sharing people's stories far outweighed the drop in income. Child custody disputes and murder trials. School board and borough assembly meetings. Local elections and natural disasters. Tales of surviving icy roads and reeling in monstrous salmon. Maintaining an unbiased approach to hotly contested topics kept me on my toes. Other subjects required that I listen but then write from my heart.

If I got the facts wrong, I reminded myself how important it was, though never pleasant, to have my errors pointed out. When I received phone calls from people unhappy with what I'd written, even when my facts were accurate, I learned the value of staying calm and listening to their concerns.

This was all heady stuff. In little more than two years after leaving Alyeska, I had gone from knocking on editors' doors hoping to sell stories to actually writing full-time. In the early summer of 2000, when I spotted an online ad for a reporter at *Today's News-Herald* in Lake Havasu, Ariz., I lost no time submitting a resume and samples of my writing. Was it possible I could return to Arizona as a writer? The thought pushed aside memories of the years of struggling to come home to Ninilchik. When offered the job, I didn't hesitate to accept.

Packing the bare essentials plus my dog and two cats into my Saturn station wagon, I said good-bye to Dad, now retired and living full-time on the homestead, to Emily, Joe and Colby, who had settled into a house only 50 miles from my cabin, and to the *Clarion*. With my friend Kathleen along to help with the move, I headed back to Arizona.

By the end of June, home was a tiny efficiency apartment. The warm water of Lake Havasu was at my feet, the desert and mountains at my back and the wonderful Arizona heat—124 degrees the morning I arrived—was everywhere. My days were spent blissfully pounding out stories about local government, feature pieces and general news. County supervisor meetings in Kingman were an opportunity for early morning drives across the desert with windows rolled down and the sun just breaking over the horizon. Coming home meant after-dark drives back to Lake Havasu beneath a star-filled sky. For a story about Ford's proving ground north of Lake Havasu I was driven around a course where cars not yet meant for the public's eyes were put through their paces. Midnight walks led by a state park ranger on the shores of the Colorado River opened my eyes to a very different side of this desert landscape.

Less than a month later came the first sign that my Arizona experience was to be short-lived. My brother Shawn sent word that Dad's health was deteriorating. He was scheduled for tests and Shawn suggested that my sister, who also lived out of Alaska, and I be present. Risa and I flew home. The tests showed what the doctor suspected: Dad had congestive heart failure.

After Mom's death, Shawn and his wife Naomi had purchased and moved into Mom's house. With them close at hand to keep an eye on Dad, Risa and I returned to our homes. A few weeks later, however, Shawn called again. Dad's health was continuing to decline, and Shawn strongly suggested I return to Ninilchik.

Ignoring Shawn's suggestion was never a consideration. If help was needed with Dad, I would be there. Kathleen flew from Alaska and my daughter Jennifer flew from Los Angeles to help me repack. Considering my mood over this sudden shift in direction, they were brave to volunteer.

Being rehired by Lori at the *Clarion* offered some much-needed stability in the months that followed. Two weeks after my return, Shawn announced he had accepted a job in Anchorage and was leaving Ninilchik. Then, in spite of doctors' recommendations, Dad insisted on following through with plans for a cruise. Accompanied by Risa and Ric, her husband at the time, Dad made a two-week roundtrip voyage between Florida and South America. By the time they returned to Risa and Ric's home in Arizona, where Dad planned to visit for a while before continuing to Ninilchik, his condition had reached new lows. Six months and three operations later, including open-heart surgery, doctors finally released him to return home.

Being well enough to travel and being well were two different things, I discovered. Dad had a long list of prescriptions to be sorted and taken at specific times, there were doctor appointments to be kept and, as time passed, there were more and more trips to the emergency room. Lori offered me the opportunity to work from home, which proved an invaluable solution to helping care for Dad while maintaining my employment.

It was a demanding time. Interviews and writing for my work were scheduled around Dad's health needs. Long and late nights were the norm. As any caregiver determined to keep her own life intact knows, exhaustion was a constant companion. The cabin and this slice of the homestead were my shelter and sustenance.

Group cookouts became a frequent and well-attended way to spend time with family and friends, and include Dad in the fun. I attacked the tall grass on the right side of the cabin to create what Colby dubbed the "pushed-over weed area." With a deck added to that side of the cabin and a small fire ring not far from the spot where the earth-warming fire had been built years before, we had the perfect setup for roasting hot dogs and marshmallows, sipping hot cider or simply banishing the chill of an evening. For the growing number of youngsters, now including Colby's sister, Sophia, a construction crew of family, friends and neighbors built a simple but sturdy platform in the branches of three birch trees, with a deck beneath.

When Jennifer and Craig made their first trip to Alaska from California after their daughter Gable was born, an outdoor gathering was the perfect place to show Gable off to family members. No matter that it was January and food was literally freezing on the picnic tables from which we'd shoveled snow. We simply put more logs on the fire and more clothes on ourselves and soaked up the blessed togetherness.

. .

SHORTLY BEFORE the end of 2001, I was informed that writing for one of the publications for which I frequently freelanced conflicted with my work at the *Clarion*. The small monthly journal was running ads from businesses the *Clarion* wanted. My byline couldn't be both places, I was told. Stop writing for the journal or stop writing for the *Clarion*. The expanding range of experiences and stories that freelancing offered and the growing number of publications for which I was writing shaped my decision to leave the paper.

For the next year I freelanced full-time, still based at the cabin. One of the publications for which I wrote with increasing frequency was the *Homer Tribune*, one of Homer's two weekly newspapers. The other, the *Homer News*, was owned by the same company as the *Clarion*.

The *Tribune* had one reporter covering hard news and free-lancers like me writing about sports, schools, the arts and other feature areas. After writing for the *Tribune* for a year and with Dad's health stabilized enough that I felt comfortable being gone from the homestead during the day, I suggested to the *Trib*'s owner, Jane Pascall, that she simply hire me instead of paying piece-by-piece. She agreed.

Writing full-time for the *Tribune* focused my attention on a new area of the Kenai Peninsula. New names, new topics, new priorities, new conflicts. After several years as a freelancer with multiple editors, it was clear to me that a strong editorial presence was lacking at the *Trib*. What it lacked, however, was made up for by the crew that developed over the two years I was there. In 2003, the paper joined the Alaska Press Club and won several of the club's annual awards.

· ·

IN 2004, ONE of the *Tribune*'s readers suggested I write a story about the U.S. Coast Guard Auxiliary's Homer flotilla. I knew nothing about the auxiliary and was eager to explore the idea. The resulting interviews included one with the flotilla's commander at the time, Sandy Mazen.

From our first conversation, I was intrigued. Born in Teller and raised in Nome, Sandy was proud of his Alaska background. Like my daughter Jennifer he had a doctorate in psychology. Retired and now living in Homer, Sandy had spent years in Arizona, where he taught at Arizona State University, set up counseling programs for inmates in the state prison system and had a private practice. He had four grown children, two of them living in Homer, and seemed to have close relationships with all of them.

Pertinent to the story of the auxiliary, Sandy had a strong connection to the sea. He was an avid sailor. With his 41-foot sailboat, the *Hanta Yo*, he had sailed from San Diego to Costa Rica and across the Pacific to the Galapagos, the Marquesas and Tahiti. He arranged for others to bring the vessel north to Hawaii, where he joined them for the final leg of the voyage, sailing it to Homer.

The day the story about the auxiliary ran in the *Trib* coincided with the flotilla's monthly meeting. Attending it with several copies of the paper tucked under my arm to give to auxiliarists was a perfect opportunity to meet Sandy in person and see if he was anything like the image I'd created during our phone conversation. Not disappointed, at the end of that night's meeting I asked Sandy how to join the auxiliary.

The first step he recommended was taking the safe boating class he happened to be teaching. Owning a boat wasn't a requirement, he assured me. The class, which met twice a week from the end of January through mid-March, deepened my interest in the auxiliary, and Sandy. At its completion, I joined the volunteer organization and was given the task of writing the flotilla's monthly newsletter.

The more I got to know about the auxiliary, the more my respect and admiration for that dedicated group of people grew. The arduous training was ongoing, and they responded whenever called to pull boaters out of harm's way. I frequently accompanied safety patrols to get photos for the newsletter. On one such patrol, with Sandy at the helm of the flotilla's 27-foot SAFEboat and Mike Riley as crew, we were returning to the Homer harbor when a column of black smoke became visible toward the center of Kachemak Bay and the short-wave radio announced a boat on fire. Sandy immediately headed us in the direction of the smoke.

Once on scene, we saw that the people from the burning boat had been removed by a good Samaritan vessel now a safe distance from the fire and were awaiting our arrival. In spite of the rolling water, Sandy kept our boat in position while Mike helped the people aboard. The first was a small child, crying and shaking with fear. Mike handed the youngster to me and I helped get the youngster and others into the SAFEboat's cabin and wrapped in blankets to ward off shock while Sandy took us to Homer.

That experience deepened my admiration for the auxiliary and for Sandy. Each time he came by the *Tribune* with suggestions for stories about the organization, I was eager to spread the word about training being offered and rescues made, and also delighted by the contact with him.

· ·

As much as my co-workers and I enjoyed our time at the *Trib* and took pride in what we were creating, the owner had other ideas. Toward the end of 2004, changes were obviously afoot. Behind-the-door arguments flared with increasing frequency. The layout person was fired. When Jane announced she was hiring "the best reporter" in the area and it turned out to be an individual she knew I'd had previous conflicts with, I took it as a sign my time at the *Tribune* had come to an end.

In the years since I'd left the *Clarion*, Lori had become the *Homer News* publisher and editor. It may be surprising that a city of roughly 5,000 residents has two newspapers, but Homer does. The *Tribune* began in 1991, first as a newsletter and then a weekly tabloid, but the *Homer News* has been putting out weekly papers since 1964.

Having made my decision to leave the *Tribune*, I emailed Lori and asked if she had any need for a no-longer-freelancing reporter.

"How can we make that happen?" was her quick response.

Two weeks later, I'd made the shift. It was thrilling to once again be working for Lori. The *News* offered a pleasant and professional work environment and strong editorial oversight. It also allowed me the opportunity to build on what I already knew and contacts I'd already made while continuing to learn about the people and events of the Homer area and the entire southern Kenai Peninsula. Driving the 40 miles between Homer and Ninilchik provided a breather at the end of each day. Being close to Dad was increasingly important as maintaining his health became more and more challenging. Usually I'd arrive home, stop at his place to say hello and see how he was doing, and he'd have the table set and dinner prepared. More often than not, the menu was his favorite: salmon in one form or another.

There were nights, though, when I'd arrive to find Dad's breathing labored, his coloring gray, his blood pressure dangerously low, and I'd load him in the car to make the 40-mile drive north to the emergency room in Soldotna, the city where his doctor practiced. On those nights, it might be well after midnight before I could get Dad into bed at home and I could crawl beneath the covers under my own nearby roof. Some trips to the ER ended with Dad being admitted to the hospital. If that was the case, I'd stay with him a day, two days, a weekend, until the doctor felt he was well enough to be released.

During the summer of 2007, Dad learned he had prostate cancer. The six weeks of daily radiation treatments ordered by his doctor were available only in Anchorage or its neighboring city of Wasilla, not on the Kenai Peninsula. It was a great relief when brother Shawn and his wife Naomi, who had settled in a Wasilla neighborhood, offered to make room in their home for Dad and to ensure he got to and from his treatments.

This meant I could remain where I was and keep working. It also meant Sandy and I could continue exploring our growing relationship. But the situation didn't last much beyond Dad's six weeks of treatments.

Shawn and Naomi announced in a phone conversation one evening that things were not going well. Having Dad with them was an unworkable situation, they said. Their solution was for me to find an apartment in Homer where Dad and I could live and I could give him the increased level of care he needed while continuing to work at the *News*.

Getting Dad settled was still the top priority. A property manager directed me to a two-bedroom apartment that was a perfect fit—ground floor, so Dad wouldn't have to climb stairs, and less than a mile from the *News*. And the homestead, and the cabin, were still within reach.

I had worried that the demands of caring for Dad would end my relationship with Sandy. Instead, the connection deepened. Sandy frequently arrived at the apartment with dinner prepared for the three of us. He gave Dad someone to visit with besides me. On days when I was particularly stressed, he'd suggest we go for an evening drive that invariably ended with me crying out my frustrations and him taking it all in.

Several months later, with Dad settled and the tensions between a father and adult daughter under the same roof mounting —and with the approval of his doctor—I began spending my nights in Ninilchik. Each morning I drove those 40 miles to Homer, stopped at Dad's to prepare breakfast, went back to the apartment for lunch, was there to prepare dinner unless Sandy had beat me to it, and then drove back to the cabin. Those nights alone were important for maintaining some balance in my life. After Sandy and I married in 2008, we spent most nights at his house in Homer and I continued to care for Dad with Sandy's support. Ocasional weekends in Ninilchik were coveted mini-vacations.

In April 2009, Risa came back to Alaska and moved in with Dad, a huge relief for me. That fall, Shawn and Naomi invited Dad to come for a one-month visit. While there, Dad had to be hospitalized and, when it was time for him to be released, Shawn and Naomi suggested he live with them full-time. Risa and I made weekend trips to Wasilla to help out in any way we could—doing laundry, housecleaning, keeping Dad company. However, the situation once again proved unworkable.

After Risa, Sandy and I had spent a Sunday visiting in Wasilla, we stopped at the cabin for the night on the way back to Homer. During the night, Risa's cell phone woke us. It was Dad. After we left Wasilla, Shawn and Naomi had driven him and all his belongings to Homer, only to find the apartment locked and no one there. Of course we immediately went to Homer. Shawn and Naomi walked Dad into the apartment and that was the end of that arrangement.

Until Dad's death in 2012, Risa carried an increasing responsibility for his care. I helped out, but it was definitely her efforts that kept him comfortable. Trips to the emergency room became more and more frequent. Hospital stays were longer and longer, with Risa and me alternating sleeping in the hide-a-bed provided in Dad's room. The hospital staff became like family, buoying us with their care and support.

Lori continued to be flexible at the *News*. I met deadlines from home, from Dad's apartment, his hospital room or wherever I needed to be. She even went the extra step and prepared meals or put together packages of groceries that helped ease the demands on Risa and me.

Through it all, time at the cabin became ever more precious. No matter the conflicts and losses life presented, this was the one constant, the place where generations of my family had lived, now infused with the energy of my children and grandchildren. It was where our history was written and our future blessed.

· ·

THE PORTION OF the homestead Dad set aside for my brother has been given by Shawn to his four children, Jason, Ember, Mercy and his youngest son, also Shawn, who, with his partner, Luna Barrows, and their daughter, Solie Jackinsky, has lived there since 2008.

"It is important that we stay here so that it remains home," my nephew told me. "It reminds the family that Ninilchik is important. . . . It acts as an anchor for my soul."

When he was four years old, Shawn and his family moved for a time into the homestead cabin I had rebuilt. He attended first grade in Ninilchik School "and some of my first, most vivid memories are here in Ninilchik," he said.

For Luna, living on the family land is an opportunity to offer daughter Solie "a discovery and appreciation of what the natural environment, the natural world can provide. How fulfilling that is. A lot of the kids growing up right now don't get dirt under their fingernails, check out the worms and bugs, ever get to appreciate and be recharged by being out in the sun."

Creating a sustainable, close-to-the-natural-world lifestyle is a commitment for this threesome.

"We have more than 120 different species of food plants that we're propagating," said Shawn, who is learning to make the most of what this place on earth can provide by talking to other Ninilchik residents, scouring historical records, taking advantage of local libraries, and "by experience, trial and error."

They have been offered an opportunity to lease their land for oil and natural gas exploration, but they have declined.

"It's not about monetary wealth, it's about true wealth," said Shawn, favoring "a permanent earth-based currency like the potato, the birch, the fireweed."

"The potato will feed you," said Luna. "You don't get very far eating paper."

After Dad's death, the three acres on which he lived in the rebuilt homestead cabin were sold. During the winter of 2015, young Shawn's sister Ember secured an agreement with the new owners that if she would remove the cabin, she could have it. After much work to prepare a site a short distance from Shawn and Luna's home, and with the help of a large gathering of people from across the Kenai Peninsula, Ember did just that. She shares her brother's interest in creating a genuine food security.

"I've always felt that the idea of 'at least do no harm' should apply to all aspects of life," Ember said when she and Shawn were interviewed for a profile of their gardening published in the *Homer News*. "What you grow and how we grow it is working in tandem with nature."[2]

Shawn and Luna dream of one day securing for the family all the pieces of the homestead that have, over the years, been sold.

"Not that we ever really lost it," Shawn said. "Others are just visitors."

· ·

MY DAUGHTER Emily, her husband, Joe, and their children, Colby and Sophia, now live in Portland, Ore., but make almost annual visits back to Alaska. In 2015, on one such visit, Colby celebrated his 19th birthday. His choice of location for the party was the yard next to my cabin, or what as a youngster he described as the "pushed-over weed area," the same site as the earth-warming I held in 1993.

Emily's homestead memories, she says, include taking a bath in the galvanized washtub in front of a wood stove, playing with her sister and cousins in the trees near Mom's house, the image of swaying branches as she pumped her legs on the pink fishing-buoy swing suspended beneath a leafy canopy, the smell of salmon smoking, the sucking vacuum-release sound of a jar of homemade strawberry jam being opened, and "the tinny sound of Auntie Grace's empty cans on a rope hitting each other as she walked to the outhouse, trying to scare off any wildlife."

Those memories are reinforced by the passing down of other generation's experiences and marked by traces found of their existence: that piece of clothing for insulation between log walls, a lone baby shoe fallen into an old well, a dilapidated treehouse. Predating our family, there are sunken areas in the ground that mark *barabaras*, dwellings of this area's indigenous people.

"I've loved sharing the property and history with my own children when we come to visit," Emily said. "Sometimes we talk about the future of the land. How will we share it with our future generations? As the (Kenai Peninsula) is disrupted, how do we keep that part of our family history alive?"

· ·

JENNIFER ALSO makes sure, when she and Craig and their two daughters, Gable and Harper, travel from California to Alaska, that time on the Ninilchik land is included. During their 2014 visit, the girls' cousin Solie proudly acted as tour guide, introducing Gable and Harper to an abundance of watermelon berries growing around the cabin.

"The experience of living in Ninilchik on our family's homestead is imprinted on me," Jennifer said. "It is not just memories, photographs or stories. It is a part of me. Like my own fingerprint. It is deeply tied to my identity and what makes me who I am. And, beyond that, it connects me with something much bigger than myself, my family history across time and generations."

· ·

IN 2001, UNOCAL Corp. was searching for natural gas reserves in a 20-mile area between Ninilchik and Clam Gulch that included 12 miles offshore and eight miles on land. Two-dimensional seismic surveys to help with that effort were conducted by Fairweather Geophysical, Veritas DGC and Kuukpik Corp. I was among those contacted by Veritas for permission to come on privately owned land to conduct the work. When I declined, the Veritas representative was surprised.

"Why?" he asked, wanting to make sure I understood that should gas be discovered, it could mean financial gain for me and my family.

My answer—made for myself, my children, grandchildren and all future generations—was simple. It's our home.

Describing the Elephant

"It's not going to be what you think it will be"

THE FLOOR AROUND MY DESK IS COVERED WITH stacks of newspaper stories, some dating back more than 80 years, all of them documenting the history of oil and gas on the Kenai Peninsula. A collection of ads, articles, brochures and other documents praises oil and gas. Along with natural gas pipeline plans and coverage of community meetings about an LNG facility are diagrams and descriptions of natural gas production, spreadsheets and maps, court documents and leases.

Books, too, in sometimes odd pairings. There's Naomi Klein's *This Changes Everything, Capitalism vs. The Climate* next to Daniel Yergin's determinedly objective *The Prize: The Epic Quest for Oil, Money and Power.* Early in her text, Klein says we can expect every victory against fossil fuel companies to be short-lived and combatted by the battle cry of "Drill, Baby, Drill." Yergin characterizes the management of carbon as a likely "contentious focus for international diplomacy in the years ahead." Wayne Leman's *Agrafena's Children: The Old Families of Ninilchik, Alaska* sits next to *Oil and Gas at Your Door? A Landowner's Guide to Oil and Gas Development*.

There's "Kachemak Bay Ecological Characterization," published on disc by the Kachemak Bay Research Reserve. Alaska Geographic's *America's Wildest Refuge: Discovering the Arctic National Wildlife Refuge*. And "Greenpeace Presents: *Deadly Neighbor—Living With Oil in Alaska*," a DVD loaned to me by Bob Shavelson, executive director of Cook Inletkeeper.

Like the parable of the blind men trying to describe an elephant, each item offers one aspect of the impact of the oil and gas industry on the world, on the United States, on Alaska, on the Kenai Peninsula, on my piece of the Jackinsky family's Ninilchik homestead. Separately, these materials are enlightening and educational, almost diverse enough to be unrelated. However, their common theme hints at a larger puzzle. One so immense it seems overwhelming. So big it threatens to drive away reason.

For perspective, Sandy and I attended the "Salem to Paris" march in support of climate justice in Salem, Ore., on Nov. 29, 2015. Sponsored by the local organization 350 Salem OR, it was one of more than a thousand events held around the world in advance of the December 2015 Paris Climate Conference, known officially as the 21st Conference of the Parties to the United Nations Framework Convention on Climate Change. Similar marches had begun two days earlier in Melbourne, Australia, where 60,000 people participated. No matter the location, the message was the same: Stop the use of fossil fuel.

The sponsor of the Salem event was related to the larger 350. org, which formed in 2008 after organizers identified climate change as the most important issue facing humankind. Its name refers to the maximum amount of carbon dioxide in the environment that is safe for humans—350 parts per million, quoting James E. Hansen, former NASA scientist and professor at Columbia University Earth Institute's Climate Science, Awareness and Solutions Program. Carbon dioxide wraps around the planet like a blanket, keeping longwave radiation from drawing heat away from Earth. The more CO_2, the more heat. Based on 50 years of observations, NOAA, the National Oceanic and Atmospheric Administration, has said it is clear CO_2 levels are rising at an accelerating rate. In October 2015, CO_2 had pushed past the 350 ppm mark and was nearing 400, according to readings taken at Mauna Loa, Hawaii, and reported by NOAA's Earth System Research Laboratory.

What is behind this continuing increase of carbon dioxide and resulting rise in temperature? The answer, say the researchers at NOAA, is primarily the burning of fossil fuels.[1]

Scientists aren't talking the kind of increased heat that comes with summer and the joy of kicking off heavy socks and warm boots for the freedom of going barefoot. When they talk about global warming, they mean heat with a capital "H." The kind that melts glaciers and raises sea levels.[2]

Arriving in Salem after the march had begun, Sandy and I had no problem finding the group as they made their way down city sidewalks. In addition to the huge "Salem to Paris Climate Justice" banner at the head of the procession, marchers' signs carried slogans such as "Support Green," "Respect the Planet," "System Change, Not Climate Change."

The gathering in Salem was modest in size, numbering maybe 100 people. Drivers honked at us. In some of the cars, people smiled and waved in encouragement. Others gave us the thumbs-down sign as they passed.

"Today marks the beginning of a journey, and we're going to have to take it together," one of the speakers told the crowd, stressing the cooperative effort needed to carry the message to world leaders meeting in Paris.

. .

TWO YEARS EARLIER, Cook Inletkeeper's Bob Shavelson was one of the first people I interviewed when work on this book began. It was planned as a book about people whose lives were being turned

upside down by the invasion of oil and gas into our neighborhoods on the southern Kenai Peninsula, I told Bob, and the idea to
delve into the subject came after hearing the divergent experiences
of two families I had interviewed while working as a *Homer News*
reporter. The topic took on personal significance, I said, when a
natural gas company wanted to lease the portion of family land in
Ninilchik to which my daughters and I hold title. Bob immediately
grasped the bigger picture, taking it beyond the confines of individual property lines.

"Climate change is the gorilla in the room. It's an embarrass-
ment that a state that relies so heavily on oil and gas is not even
having a discussion about it," Bob said, linking Alaska's avoidance
of the topic to a "jobs-at-all-cost mentality and the dichotomy created that we have to make a choice between jobs and the
environment."

His comment called to mind a brochure I'd seen in the lobby
of the Homer branch of First National Bank Alaska entitled
"There's a good chance your job depends on petroleum." It
pointed out that one third of all Alaska jobs is directly connected
to petroleum. According to the Alaska Oil and Gas Association,
that one third equals 110,000 jobs throughout the state.

"We believe that's a false choice," Bob said about jobs versus
the environment. "If you don't have clean water and air, you don't
have an economy."

As Cook Inletkeeper's executive director, Bob had a front-row
seat on the nonprofit organization's struggle to guarantee clean
water, healthy fish and wildlife and strong communities, as well as
clean energy and lasting jobs. It also meant he saw "this slow-
motion bulldozer ripping apart our democracy and fouling the
only planet we have."

His hope for change came from looking to home and family
and "the fact that no one can predict the future. If you just keep

pecking and pecking and pecking, maybe something will happen. . . . Hopefully global warming won't get to cataclysmic proportions before we wake up, before we realize BP and ConocoPhillips and Exxon don't have our best interests in mind."

Former New Yorker Linda Feiler helped give Cook Inletkeeper its start in 1994. Linda had come to Alaska in 1977 eager to use knowledge learned other places that could help Alaska "get it right." In short order, she became known as "that fucking environmentalist," she said, laughing at the title she was proud to have earned. She's carried that high level of enthusiasm into public meetings like one held by Hilcorp Alaska that she told me about.

"They said, 'We're here to tell you we're your environmental oil company. We care about wildlife and are determined to convince you we are going to be working with you.' So, I raised my hand and said, 'Is that why you got a permit to get your gravel from the bottom of salmon streams? Why you're doing seismic testing and blasting right in direct line of where sleeping bears are at the moment? Why you're cutting across the wildlife refuge?'"

When she attempted to ask questions at another meeting, she was cut off.

"They said, 'We're not here to answer questions,'" Linda recalled.

She didn't mince words in describing her frustration with the oil companies in Alaska. Of the *Exxon Valdez* spill, she exclaimed, "You (Exxon) have lied, cheated, destroyed, maimed, mutilated how many things without a license? How many hundreds of thousands of animals? One guy goes out hunting at the wrong time and he's in prison for five years. And you have these guys, every single one, that should be jailed for life."

Recollections of the *Exxon Valdez* spill still haunt her, particularly the experience of one friend who worked on a beach cleanup crew, amid animals so covered with oil as to be unrecognizable.

"She came home hysterical, crying because she couldn't get it out of her mind that when she was jumping over the rocks, one of the rocks cried out in pain. It was a seal," Linda said.

In spite of her frustrations, Linda revealed a less feisty side when she spoke of her deep affection for the planet. Summing up, she drew from the words of Dante Alighieri: "Nature is the art of God."

. .

GREG ENCELEWSKI is the president of Ninilchik Natives Association Inc. Like me, he is a descendant of Ninilchik's founding families and grew up in the village. Also like me, Greg worked in Alaska's oil patch. His employment on the North Slope began in 1983; he eventually transferred to work in Anchorage and then retired in 2009.

"NNAI has 69,000 acres in the Ninilchik area and some of that acreage has leases and oil and gas development," he said. With surface rights to the land, "we interact anywhere there's surface development on our property."

An increase in people in the Ninilchik area and more traffic are some of the changes he has seen as a result of oil and gas activity. As to the availability of more jobs, "that's pretty well tied up by the people in the industry. . . . Most of these jobs they have, as far as once the (exploration and drilling) work is completed, the operator jobs come from a workforce they already have, so there are very few high-paying jobs for locals. Or long-term jobs."

Although jobs for local people may be temporary, Greg maintained that the overall oil and gas work does benefit local businesses, such as restaurants and lodging. Combined with other forces, however, it was creating change with an uncertain outcome.

"Whether we're born and raised here or homesteaded here, we live here for a lifestyle that is rapidly changing. Changing in a lot of ways. It's not just oil and gas. We've lost fishing, clams, crab, moose. All of them are dwindling. That's in our lifetime. What's going to happen?" he wondered. "Oil and gas, as it creeps in here, is not so bad, but then we get caught in the development. It's changing (the peninsula)."

. .

TERRY REID ran headfirst into the surface rights NNAI holds on its land. A retired U.S. Forest Service employee, Terry moved to Ninilchik in the late 1990s. He'd been visiting from Anchorage more and more frequently to fish when the idea struck him that he could buy a piece of land and build a garage where he could store his boat, rather than towing it 185 miles each way. A friend told him about an available three-acre parcel owned by NNAI.

"They had a provision in there basically for unimpeded access to the site for any kind of exploration or wells or whatever," Terry said. "I saw that, and I'm a little bit of a lands person—it's part of my background—and I said no, full access without cooperation would not work for me, certainly not on three acres of land and probably not even on 50 acres of land."

Through negotiations "we finally agreed that they would not have unfettered access to the land," Terry said. "That means (oil or gas companies) can do directional drilling or whatever they want and suck out all of the oil. It just means from the little negotiation and little squabble we had that it would be very difficult for them to have access over my lands with a road or pad. I think a lot of people take the deed and sign it. I just didn't want the risk of someone putting in a drill rig."

. .

BRUCE OSKOLKOFF is another descendant of Ninilchik's founding families. He has worked for or with Ninilchik Natives Association Inc. for forty years, dealing mostly with land and resource issues. In addition to his role with NNAI, Bruce has a family connection to natural gas. In 2002, Unocal Alaska announced that the

Grassim Oskolkoff No. 1 well, named for Bruce's now-deceased father, was the first exploration well drilled under a joint operating agreement between Unocal and Marathon Oil in the 25,000-acre Ninilchik exploration unit.[3] Marathon estimated the well contained 60 billion cubic feet of gross proven recoverable gas reserves. The 14-acre pad now includes seven wells, according to Bruce, and is operated by Hilcorp Alaska LLC.

"It's right there on my mom's property and I manage all that for her. I wrote all those contracts," he said.

"The thing I've learned," he continued, "is there's a whole bunch of fear when you aren't well educated about that line of business. But the more you learn, the clearer the understanding you have. . . . One thing I could never get away from is I like my cars, I drive to work, I have a motorcycle. I have to put myself in the same group of people that want some benefit, but realize the risk. . . . I just feel that instead of taking a hard-line view, I try to look at what's really the big picture."

Bruce's brother, Gary Oskolkoff, talked about his interaction with oil companies in a segment of the 1994 30-minute Greenpeace film *Deadly Neighbor: Living With Oil in Alaska*. In it, Gary directed viewers' attention to drilling residue mud in an abandoned waste site he stumbled into while walking through the woods near his home.

"This particular chunk of drilling mud is fairly upsetting to me because of the fact that I did fall into it," he said in the film, showing the gray muck he shoveled out of the ground for the camera. With the mud's composition unknown, Gary said, he had been concerned when he suffered irritation to the skin on his hands after the fall. "I'm pretty disgusted on a personal level. You can't help, after you've had first-hand experience in these woods as I have over the years, to be disgusted with those who tend to leave their garbage—and that's basically what the oil company's done.

It has come here, it has taken what it's wanted, and it has left its garbage."

. .

A NINILCHIK resident since 1975, Steve Vanek has had dealings with oil and gas companies on multiple fronts. He was offered a lease on some property he owns north of Ninilchik but "said no because they didn't want to pay anything for it." When approached about having seismic work done on his property in the village, Steve again said no.

"It's not right they want to use your property for their gain and not reimburse you. They didn't want to pay anything," he said. "I'm not against development, because it does help with the economy, but when it's done cheating-wise, I don't think it's right."

As the president of Ninilchik Emergency Services' board of directors and one of the first trained EMTs in Ninilchik, however, Steve has accepted financial donations from oil and gas companies working in the area that benefit the emergency response organization.

"They expect us to be covering them in an emergency, whether it's fire or an ambulance emergency. It's to their benefit to have us here, so I see that as a positive thing for them as well as us," he said.

Steve is also a fisherman. If a drilling rig were to be planted in the middle of prime fishing areas of Cook Inlet, he said, "we wouldn't be too happy about that."

"We do have drill rigs in the northern Cook Inlet area," he said, "and I've fished up there one or two times and people seem to be able to avoid them all right. The one they had at Stariski— we really don't fish down there so it didn't bother us."

Steve is part of a local group of fishermen and boat owners who take oil-spill-response training with industry personnel and the U.S. Coast Guard, an activity he assigns a plus and a minus.

"I get anywhere from \$700–\$800 for two to three days (of training). It's a good thing to do and I don't want to be putting the kibosh on it, but I really feel that if it was a rough day out on the water, what we're doing isn't going to help," he said.

I asked Steve if he thought the oil and gas industry was making overall positive or negative impacts on the Ninilchik area.

"Those kinds of things aren't going to happen overnight," he said—and then added, "They sure came in and knocked Emil Bartolowits' place down and flattened (it) overnight. So maybe it can happen overnight!"

. .

MIKE SCHUSTER is a construction subcontractor. After moving to Ninilchik full-time in 1995, he began offering overnight accommodations to visitors. Located on Oil Well Road, Meander In B&B overlooks Ninilchik River as the river winds its way toward Cook Inlet. From Mike's living room mama brown bears can be seen in the summer with their cubs, foraging in the river valley. For him, the presence of oil and gas crews in Ninilchik is "a show-me-the-money deal. I think money obscures the issues. It hides the real thing."

The "real thing" would be the difference he sees oil and gas having on the area.

"What defined the Kenai Peninsula is that we're the playground for Alaska and a large percentage of the visitors to the state. It's the natural beauty and the fact that we're not urban. We're very rural. The communities are small and unique. We have a history and we have rich resources, and people here choose to live where the resources are immediately available," he said. "Then you have something like oil and gas development, external structures, gas lines buried, substations, distribution stations. . . . There is no way that gas and oil development on the Kenai Peninsula as it is, for the next ten to fifteen years, can do anything but destroy the aesthetics of living here."

When he's expressed those views, he said, friends have said, "Oh, there's going to be more development, more people moving down here, more people participating in the economy. Hang in there, Mike, things will get a lot better because of that." But Mike is not so easily swayed. He compared what's happening on the Kenai Peninsula to what he has observed in his home state of Texas.

"It's the grossest thing in the world. There's a big 'man camp' (industry term for worker housing) on the highway. There's an abandoned one three or four years old. There's burning off wells. It's just all about the money, and the only people benefiting really are the ones making the money," he said. "That's what I saw in Texas and that's why I don't buy into it here. I'm advising everyone to be real conservative. Don't think it's a big opportunity. It's not going to be what you think it will be, and the consequence is that this will no longer be the beautiful place everyone wants to come to."

. .

MILLI MARTIN, former member of the Kenai Peninsula Borough School Board as well as the Kenai Peninsula Borough Assembly, representing the southern Kenai Peninsula, speaks of the need for zoning and the challenge that concept poses for independent-minded Alaskans.

"You don't expect commercial development in the middle of residential areas, but, unfortunately, the way the Kenai Peninsula has developed, we don't have zoning. We have practically no protection for neighborhoods unless they are aware of (the borough's) local option zoning they can avail themselves of," she said. "They don't realize that oil and gas or anything else could come into their neighborhood because they're not protected."

Milli is personally aware of that threat. When she and her now ex-husband bought their 160 acres, a title search revealed "the original homesteaders, possibly needing money, sold the rights to Chevron." At the time of the purchase, Milli was focused on saving a piece that size from development.

"The idea that Chevron could come in and do what they want never occurred to me until some years later," she said.

Knowing she has no control over what can happen below the surface of that piece of properly is a situation Milli described as "scary. . . . That the previous owners sold the subsurface rights sure complicates my life now. It's a dark shadow," she said. Those concerns created a different scenario when additional acreage was purchased later.

. .

YEARS BEFORE Milli came to Alaska, Yule Kilcher was finding his way to the southern Kenai Peninsula from Switzerland. He arrived in 1937, homesteaded east of Homer, and with his wife, Ruth, raised eight children on the shores of Kachemak Bay. He served in the Alaska Senate from 1963–1967 and died at the age of 85 in 1998.

"He was conservation-minded and (favored) using the resources wisely," his daughter Mossy said. "And he basically felt oil and gas belonged to Alaska. After all, why go through all the trouble of a constitution if it didn't mean we had more control over it? But he saw the writing on the wall. He saw that they were going to try every trick in the book to see what they could get out of here and leave as little money or oil behind as possible for us and for Alaska. He made no bones about it. He spoke his mind."

In one instance Sen. Kilcher thought his outspokenness had sparked an attack.

"I don't know the year," Mossy recalled. "It wasn't during the time he was in the Legislature, but he was at a meeting in Seward, at a hearing, I think—one of those oil hearings. He was ranting and raving at that meeting, thinking it was a sell-out. On his way back, a tree fell across the road right in front of him and the car crashed into it."

Her father was alone when it happened. His vehicle was dented; he had a bloody nose and was taken to the hospital.

"Later when he went back, the tree had been cleared, but there was a freshly sawed stump right beside the road. He was convinced someone had been sent out to get him. And you know, it could have been. We'll never know," she said.

Hearing Mossy's story and her father's suspicions didn't surprise me. In my meetings with people in the past few years, I've met several willing to share their experiences with the oil and gas industry but unwilling to have their names used. They refused photographs and wanted details blurred so they wouldn't be identified. They worried about repercussions from pro-oil and pro-gas neighbors and industry officials. They worried about the welfare of their children and grandchildren. Some were bound by signed contracts that forbade them from divulging any information. Unless they said differently, I had no choice but to honor their concerns.

The Salem to Paris march that Sandy and I participated in closed with the singing of "We Shall Overcome," the anthem of the civil rights movement of the 1960s that now has been sung worldwide. Leading it, musician Peter Bergel directed us to sing a "We are not afraid" verse and I was mindful of each of those people who had cautiously shared their stories if I promised not to use them. Perhaps one day we will all sing fearlessly the verse that insists "The truth shall make us free."

. .

As a 26-year board member of Kachemak Bay Conservation Society, Roberta Highland has been involved in oil and gas discussions and controversies for "going on probably 30 years," she told me.

"For the sake of future generations, we've got to get this fossil-fuel train wreck stopped. I can't get over that our state never talks about climate change, ocean acidification and global warming. They don't even acknowledge that it's there," she said.

Like Bob Shavelson with Cook Inletkeeper, Roberta is committed to staying hopeful.

"I do think that we'll figure this out, but it's going to be kicking and screaming the whole way. I do know that environmental conservation groups are absolutely on the right track. There's just no question in that," she said. "We have a beautiful place. We owe our land and Mother Earth in this state, in this area, we owe it to future generations, to protect it and be good stewards."

· ·

WHEN I SPOKE with Homer author and fisherman Nancy Lord, she addressed the importance of planning and zoning, "negative words in Alaska. . . . If there are places where we want to have a rural lifestyle, that's probably not the best place to have industrial development. In general, we just need to be aware you can't do everything everywhere. We have to be regulating more than we are."

Beyond regulations, however, Nancy stressed the need for vigilance.

"These things are incremental, the wearing away. Small pieces at a time until what we value is gone. So vigilance, I think, and not letting the Legislature, for example, water down our protections. We need to do a better job of protecting habitat and water," she said.

In her book *Early Warming, Crisis and Response in the Climate-Changed North*, Nancy details a wearing away that is occurring. Her book opens with a look at the Kenai Peninsula and the efforts of local scientists to safeguard the future by understanding what has happened in the past and is happening now.[4] Higher temperatures allowed spruce bark beetles to survive winters, destroy four million acres of spruce forest and rob streams of crucial salmon habitat. Similarly, invasive plants capable of turning streams into marshes and previously unable to survive Alaska's harsh climate have been found thriving in 259 locations including salmon streams. Between 1968 and 2009 local meteorological records reflect a 60 percent decline in available water in the Kenai lowlands. On a local level, less wetlands means less natural firebreaks. On a global climate level, diminishing wetlands play havoc with the cycling of carbon dioxide and methane.

Lord doesn't stop there. *Early Warming* reaches beyond the Kenai Peninsula into other northern regions in an attempt to unravel the tangled web of science, politics and life on Earth past, present and future. In her introduction, she refers to a statement that she keeps on her desk, by John Holdren, President Obama's assistant for science and technology who is also director of the White House Office of Science and Technology Policy and co-chair of the President's Council of Advisors on Science and Technology. "We basically have three choices," Holdren had said. "Mitigation, adaptation, and suffering. We're going to do some of each. The question is what the mix is going to be. The more mitigation we do, the less adaptation will be required and the less suffering there will be." Key components, Holdren and his co-chairs wrote in a 2013 report to the president, include decarbonizing the economy, leveling the playing field for clean-energy and energy-efficient technologies, and sustaining research on next-generation clean-energy technologies.[5]

Making the point that Alaska is Ground Zero when it comes to climate change, Lord says in her book, "Maybe we'll show how to turn a crisis into stronger communities and a more sustainable future. Or maybe the lessons will be quite different; maybe what we'll learn is how hard it is to lose homes and livelihoods, the costs of ignoring risk and peril, what it means to suffer."

RETURNING TO where the journey for this book began, I checked in with the McConnells and Paulsruds on North Fork Road.

The McConnells were the ones with the "Drill, Baby, Drill" sign in their front yard, echoing the line made famous by Alaska's former Gov. Sarah Palin. (In her Jan. 19, 2016, speech endorsing Republican presidential hopeful Donald Trump, Palin issued a right-all-the-nation's-wrongs version of familiar rhetoric: "Well, then, we're talking about our very existence so no, we're not going to chill. In fact it's time to drill, baby, drill down, and hold these folks accountable."[6]) Mark and Juley McConnell still live across the street from the well pad. Mark is now employed as an engineer aboard a tugboat, "a good job that he doesn't want to leave," Juley said. With their children growing up, including one offspring attending college and another working on the North Slope, the couple are considering selling their house for something better suited to the family's shrinking size and Juley's expanding specialty embroidery business.

On a trip to Georgia after the death of Mark's mother, Juley came across a newspaper clipping that included a photo of their house featuring the "Drill, Baby . . . " sign in the front yard.

"I thought, oh, my gosh, all the way down here? Don't tell me!" she said, laughing.

Rick and Lori Paulsrud continue to live on their 3.5 acres next door to the well pad.

"We've contemplated moving a couple of times, but we haven't tried to put it on the market. Most people that are going to want to be more rural don't want to be next to noise and lights. And there's *constant* noise. . . . I can't imagine anyone willingly buying the house," Lori said.

"We always liked our little place," Rick said. "This just puts a big damper on it, having that well pad next door."

I asked Lori where she'd choose to move. "I guess . . . maybe a location where I can do the things we're doing now—ski, snowshoe, garden, have critters—without having a natural gas well in the backyard."

· ·

ON ALASKA'S Kenai Peninsula, where oil and gas exploration and production are making their way into more and more backyards, where neighboring drill rigs tower above ground while underground they extend their reach in multiple directions, protecting *home* becomes vital. And in a world where the temperature continues its dangerous increase as we stubbornly hold on to our oil and gas addiction, the definition of "home" expands to global proportions, taking on intimidating dimensions.

"People feel that they're not important. They've been made to feel that they don't count when in fact they're the only thing that counts," said Jim Hightower, former Texas agriculture commissioner, in the award-winning documentary *Fuel*. "The genius of America is not in big corporate power. It's not in big government. It's in people who understand this notion of the common good."[7]

Anchorage author-journalist-mountaineer Art Davidson gave that thought an Alaska perspective.

". . . I'd like Alaskans, myself included, to take more time to appreciate how fortunate we are to live in one of the most spectacular and abundant natural environments on earth. This land is good to us—we have to find ways to be more responsible for it. The process of change starts with one person. And it starts with a personal decision."[8]

In other words, as Sandy and I were reminded at the Salem to Paris march, "If we can do one little thing, we can do a lot of things."

Three acres is little compared to the rest of the Kenai Peninsula. To Alaska. To the planet. But when Hilcorp's offer for a lease to have their way with the little piece of family homestead entrusted to my daughters and me arrived in the mail in 2013, it didn't feel little. Not the offer. Not the impact it would make on my family's life.

The offer came at about the same time Hilcorp was facing fines by the Alaska Oil and Gas Conservation Commission (AOGCC) for failing to test blow-out equipment at their operations in the Swanson River field, just one of "more than a dozen enforcement actions" aimed at the company. An *Alaska Dispatch News* story quoted from AOGCC's order: "The aggressiveness with which Hilcorp is moving forward with operations appears to be contributing to regulatory compliance issues."[9]

Two years later the *Houston Chronicle* and *Fortune* magazine would run stories of Hilcorp employees receiving $100,000 bonuses for helping the company double its size in the span of five years.[10] In December 2015 *Alaska Dispatch News* reported that Hilcorp was being fined again by AOGCC for the near-deaths by suffocation of three North Slope workers at the Milne Point oilfield.[11]

Noting that the company was facing other fines because of compliance failures, the newspaper quoted once more from the AOGCC: "The disregard for regulatory compliance is endemic to Hilcorp's approach to its Alaska operations and virtually assured the occurrence of the incident" at Milne Point.

· ·

THE ARRIVAL OF the lease offer in 2013 came before all that. It was just one company making one request for one piece of land, with one decision to be made.

For my daughters and me, "one little thing" was not signing the lease.

This Song Is Not My Own

By Randall Williams, from the CD "One Night in Louisiana,"
Musafir Music, 2007—Randall Williams (BMI)

This song is not my own

What began as an egg will fly like a bird

It will speak to you in ways that I'll never know

And you'll hear stories that I never told

It's the way of the world and its riddles untold

Maybe there's God or not I don't know

This song is not my own

This life is not my own

Cousteau says I can't have it alone

The paths that I have walked, the songs that I've sung

Like somebody gave me a job that I've done

It's the way of the world and its riddles untold

Maybe there's God or not I don't know

This life is not my own

This world is not our own

We worship at the feet of the unknown

Holding each other's hands we walk down the road

A comfort in the dark night of our souls

It's the way of the world and its riddles untold

Maybe there's God or not I don't know

This world is not our own

This song is not my own

I'd give you an egg but it'll fly like a bird

It'll speak to you in ways that I'll never know

You'll hear stories that I never told

It's the way of the world and its riddles untold

Maybe there's God or not I don't know

This world is not our own

This life is not my own

This song is not my own

JIM

"They're planning on fracking everything.
That doesn't give me a warm fuzzy feeling"

I T'S BEEN A STRUGGLE, WITH THIS VOLATILE TOPIC, to know
when to stop scheduling interviews, quit poring over reference
materials, and let go. There were individuals I wanted to talk
with, daily headlines to raise eyebrows and pique interest, announcements of plans for relevant activities. But the book was finally being readied for the printer in late March 2016 when I
spoke with Jim Arndt, a southern Kenai Peninsula resident who
had become aware that the mineral rights on his 80 acres of agricultural land east of Homer had been leased to Hilcorp Alaska.
What I learned from Jim and his wife, Mary, raised other questions. By then, it was so close to the Legislature's April 17 adjournment that it was worth the wait to see what the Senate and House
of Representatives might do about the state's $4 billion budget cri-
sis, brought on by heavy dependence on oil and gas revenues.
What occurred in those final few weeks underscores that Alaska's
love affair with the oil and gas industry and the long-reaching impacts of that affair are far from over.

AFTER 26 YEARS in Fairbanks, a landlocked city in Alaska's heartland with winter temperatures known to dip into the minus-60s, Jim and Mary Arndt decided to search out "an opportunity for a life change." Two years later, the couple and their sled dogs settled on a 40-acre parcel east of Homer. It was classified by the state as agricultural land, a perpetual covenant restricting the land to agricultural purposes and specifying subdivision restrictions, and was located more than a bit off the beaten path.

"We had to cut our way in," Jim said of improvements he and Mary made to the four-wheeler trail accessing the property, a trail that continues into the Kenai Peninsula's backcountry and is sometimes used by hunters, snowmachiners and other outdoor enthusiasts. "It's pretty remote, but that's the way we wanted it."

The Arndts later purchased the neighboring 40-acre parcel of agricultural land and planted 30 acres of Timothy hay. For 10 years they enjoyed the quiet setting of what Jim described as his and Mary's "semi-retirement home out on the edge of the country."

What the Arndts didn't fully understand at the time they purchased the original property was that the state of Alaska maintains the mineral rights on land designated "agricultural." In fact, they were to discover, their 80 acres already holds an old oil well that according to Jim had been abandoned in 1979.

Then, in 2015, Hilcorp Alaska successfully obtained oil and gas leases in the area, and in early 2016 the company began seismic exploration of what lay beneath the Arndts' feet.

"They (Hilcorp) wanted us to sign a surface-use agreement to do seismic work. I didn't sign it, but they can do whatever they want. We can't deny them access," Jim told me in April 2016. Hilcorp's access is made easy by the 60-foot wide public easement on which the trail cuts across the Arndt property. As we talked by phone, Jim said he could look out his window and see the work going on.

. .

NEXT DOOR TO the Arndts are Chrisie Trujillo and Chris Hutson. Although they have not been approached with a lease offer by Hilcorp, Chrisie and Chris are dealing with the intrusion on their land that is occurring as a result of their close proximity to the Arndts.

"Our complaint is that they're not (working) on our property, but they're trying to get to Jim and Mary's land, so they come down our driveway thinking it's the road, which it isn't," Chrisie told me by phone, of the use of her driveway by seismic crews from Hilcorp's contractor, Global Geophysical Services. While most of the driveway runs along a section line easement that allows for public access, about 600 feet of the drive in to their home is outside the easement. Added to that is a seeming lack of communication among the seismic crews. The workers told to stay off the driveway one day may not be the ones who show up to work the next day, making it necessary for Chrisie and Chris to repeatedly warn them away. Chrisie is hoping for better results after placing a sawhorse and a "No Trespassing" sign across the driveway.

Compounding the situation is the time of year: spring, or, as Alaskans refer to it, breakup, when snow and ice melt and what appeared as solid ground turns into mud with a capital M. During breakup, driveways become deeply rutted tracks that bog down and sometimes seriously damage vehicles. After spending hours filling in resulting holes, smoothing out rough spots and repairing those vehicles, property owners learn the importance of waiting for the muck to dry in summer's warmer days.

"We've spent the last nine years getting our road in, but these guys don't know anything about breakup. The nice thing would be if they said, 'We'll give you a couple loads of gravel for messing up your road.' But we'd already told them we didn't want them out here," Chrisie said.

On Easter Sunday it wasn't traffic on the driveway but the noise of helicopters delivering seismic equipment that disturbed the quiet of this remote neighborhood—by Chrisie's count, 30 to 40 trips made in and out during the day.

"It's frustrating knowing someone can come into your land and just take it over if they want to," Chrisie said. "I'm pro-development, but in someone's back yard? It feels like some old Western movie where the railroad wants to come through the land you've had for five generations and they'll come through whether you want them to or not. I'm just hoping they don't find anything."

. .

How ARE PROPERTY owners notified when mineral rights are leased for the land on which they own the surface rights, I asked Marta Mueller, natural resource manager of lease sales for the Alaska Division of Oil and Gas. She referred me to Alaska Statute 38.05.945, which spells out notification requirements for the Department of Natural Resources, of which Oil and Gas is a division.

"That prescribes how we communicate with people that there's going to be an upcoming lease sale," she told me by phone. "It includes notices to postmasters, notices to libraries, notices to newspapers, notices posted online—and all Alaskans can subscribe to daily notices. Should exploration be done on a leased area, then it goes through another permitting process with a different set of requirements for notification with opportunities for the public and stakeholders to give feedback along the way."

But are individual property owners notified, I asked, specifically thinking of the Arndts. No, Marta said, reminding me that lease sales typically include multiple tracts totaling millions of acres.

Diane Hunt, special projects and external relations coordinator for the division, referred me to Alaska Administrative Code 11 AAC 96, "Miscellaneous Land Use" (regulations for the Department of Natural Resources), to answer questions about permitting notifications. With regard to the seismic work occurring on the Arndts' land, she said, public notice of the work and an invitation for the public to comment appeared on Feb. 9, 2016, in the *Alaska Dispatch News*, the state's largest newspaper, published in Anchorage, and the *Peninsula Clarion*, published in Kenai, as well as in the *Homer Tribune*, one of Homer's two weekly papers, on February 10. The deadline for the comment period was March 9. The public notice also had been distributed to postmasters in Homer, Anchor Point, Nikolaevsk, Ninilchik, Kasilof, Kenai and Soldotna.

"In addition, the division encourages public outreach to communities, and Hilcorp provided an outreach meeting in Anchor Point on February 24," Diane said of a meeting held in Homer's neighboring community, a driving distance of approximately 20 miles west and 15 miles north from the Arndts.

In response to my question about whether leases or permits have ever been denied or canceled because of a property owner's concerns, Diane said no. "However," she said, "it can impact mitigation measures, so, again, the public has an opportunity to express concerns and we have to respond to those concerns."

Land use conflicts that arise are "most often" resolved between the applicant and interested parties, she continued. When a resolution is not forthcoming, the division steps in.

"In the event a resolution cannot be found, it is most common for the applicant to withdraw the application," Diane said.

The state has amended leases and partially refunded them when a title conflict has been discovered, but lease amendments are rare, she said. The only instance she could recall of the state buying back leases occurred in 1977 under Gov. Jay Hammond, after it was determined that inadequate public notice had been given for a lease sale that included tracts in Kachemak Bay.

Regarding limitations to Hilcorp's surface development on the Arndts' agricultural land, the answer seemed to be difficult to sum up.

"There are many limiting factors for surface impact," Diane told me by phone in response to my emailed questions. Mitigation measures, project-specific concerns, use of hazardous materials, discharges and the proximity of habitat, how surface, subsurface and mineral ownership are divided are just some of the many aspects managed by the division and a host of other entities.

"There's a variety of packages in the state of how things are set up," she said.

. .

AT THE BEGINNING of the year, the *Alaska Dispatch News* carried an announcement from BP that because of falling oil prices the company would be cutting 4,000 jobs worldwide and an estimated 210 jobs within the state.[1] A few months later, another announcement tied to worldwide oil prices: BP planned to reduce the number of rigs it operated at Prudhoe Bay.[2] Almost simultaneously, Apache Corp. announced that it was shutting down Alaska operations.

"Due to the current downturn, Apache has had to significantly scale back operations and spending," said a statement in the *Alaska Dispatch News* from Apache's public affairs office. "We recently reduced our spending plans for 2016 by 60 percent from 2015 levels and are focusing our limited dollars on specific international opportunities and strategic testing in North America."[3]

The measure of Alaska's dependence on oil and gas revenues is painfully evident in the state's $4 billion budget gap, which as noted earlier has grown as revenue received from oil and gas production has declined. Low oil prices combined with the state's system of tax credits—allowing oil and gas companies to write off operating losses and efforts to encourage new development—have resulted in the state currently paying the companies nearly as much in tax credits as it receives from industry revenue.[4] Gov. Bill Walker has offered a multifaceted remedy for the state's fiscal crisis that includes increasing oil industry taxes, but state legislators are apparently reluctant to take such a step. As April 17 loomed over the 29th Legislature, marking the end of the 90-day session limit approved by a voter initiative in 2006, Walker threatened to call a special session if the 60-member body failed to pass any of his tax proposals. (Kara Moriarty, president and CEO of the Alaska Oil and Gas Association, has been quoted as arguing that the industry is "not some cash cow. We don't have pools of money sitting around in reserves in this low price environment."[5])

On April 16, Senate President Kevin Meyer, a Republican from Anchorage who is an investment recovery coordinator for ConocoPhillips and has served in the Senate since 2008, didn't expect the Legislature to make the 90-day deadline, because of disagreement over the state's oil and gas tax credit system. After meeting with Gov. Walker and House Speaker Mike Chenault, a Republican from the central Kenai Peninsula community of Nikiski, Meyer said the three agreed to a plan for the Legislature to continue working an additional week or two and, if no compromise could be reached by then, continue working in Juneau or possibly Anchorage.[6] Shortly after midnight on April 17, the *Juneau Empire* posted online an update saying that, while hoping for an on-time adjournment, Gov. Walker was still considering calling a special session if differences on the state's budget problems weren't resolved. The story also reported Chenault saying, "I'm willing to give a little bit, but I think we need to be very careful with the changes we make to the tax structure."[7]

The 90-day deadline came and went without the Senate and House reaching agreement on the state's operating and capital budgets or on Walker's proposals to rewrite the state's oil tax credit subsidy program, reinstate a personal income tax that was repealed in 1980 when the state was awash in oil and gas dollars, and restructure the Permanent Fund to help pay for state government.[8]

In 2006, the 90-day-limit ballot measure had passed with 117,675 or 50.83 percent favoring the limit and 113,832 or 49.17 percent voting against it.[9] The constitutional deadline is 121 days.

As discussed previously, when Senate Bill 21, a move to cut oil taxes, was making its way in 2013 through the Legislature, conflict of interest concerns were raised regarding Meyer and Sen. Peter Micciche, a Republican from the Kenai Peninsula city of Soldotna who also is employed by ConocoPhillips. When the bill came before the Senate on March 19, 2013, Meyer and Micciche asked to be recused from voting. According to the Legislature's Uniform Rules, however, a recusal requires unanimous consent. The Senate Journal for that day notes "objections were heard" for each senator's request, and Meyer and Micciche were "required to vote."[10] Two days later, the bill passed the Senate by two votes, 11 to 9.[11]

In an *Alaska Dispatch News* article addressing lawmakers' potential conflicts of interest, Micciche is reported as having requested a formal advisory opinion from the Legislative Ethics Committee, with no conflict found.

"There was not an actual conflict in the case of SB 21, but I can understand where the perception of a conflict could exist for folks unfamiliar with the (Ethics) Act," Micciche wrote in an email quoted in the article.[12]

Meyer argued, according to the article, that not voting would have left his 70,000 constituents without a voice.

"You'd have a population the size of Fairbanks being totally disenfranchised down in Juneau," he said.

Former Alaska attorney general John Havelock also is quoted in the article, "It's outrageous that they should be required to vote," he said.

. .

ACCOMPANYING ALL the legislative and corporate turmoil are national and international stories on climate change, now widely acknowledged to be linked to the burning of fossil fuels. In March 2016, KTUU, an NBC affiliate in Anchorage, broadcast a brief report on one way in which climate change is affecting Alaskans. In the Bering Sea community of Unalakleet, sea ice that decades ago was several feet thick is now a sheet of mere inches, a grave concern to subsistence hunters trying to feed their families.[13] In April 2016, the University of Alaska Fairbanks and the U.S. Geological Survey released a study indicating that rising temperatures since 1860 have resulted in not only bigger shrubs, but shrubs in a wider range, which in turn has made it possible for shrub-eating moose to extend their habitat from forest to tundra. "These northward range shifts are a bellwether for other boreal species and their associated predators," the study says.[14]

. .

BUT THE RESEARCH and the reporting look farther back.

A 2015 investigative series carried by the *Los Angeles Times* and other newspapers reported that a senior researcher for an Exxon subsidiary told a conference audience in 1991 that greenhouse gases were rising "due to the burning of fossil fuels. Nobody disputes this fact."[15] In 1996, the series reported, Mobil Oil planned for rising temperatures and sea level when it designed facilities for the Nova Scotia coast. "An estimated rise in water level, due to global warming, of .05 meters may be assumed," Mobil engineers wrote in design specifications for the Sable gas field. As

Exxon, Mobil and Shell were fending off regulations to address climate change, the joint investigative team from Columbia University's Energy & Environmental Reporting Project and the *Los Angeles Times* found, the companies were raising the decks of offshore platforms, protecting pipelines from increased coastal erosion, and including Arctic warming and buckling concerns in their design of helipads, pipelines and roads. Dismissing questions over this apparent contradiction, oil industry representatives say it is simply good business.[16]

. .

AND LOCALLY? As BP slashes jobs and reduces operations, as Apache packs its bags and heads elsewhere, as the governor threatens and legislators drag their heels, as oil industry advocates claim empty pockets, and with the threat from climate change virtually unquestioned, activity on Cook Inlet and the Kenai Peninsula charges ahead. A jack-up rig tied to a city of Homer dock is being prepared to move into location at Furie Operating Alaska's Kitchen Lights Unit in the upper Cook Inlet. SAExploration is gathering 2D and 3D seismic data for an area near the central-peninsula community of Nikiski. And BlueCrest Energy has announced fracking plans for its Stariski wells on the Hansen pad, with the help of a 15-story, $77 million rig being shipped from Texas and scheduled to begin drilling in late June. Described by Ben Johnson, BlueCrest Energy's chief executive, as the most powerful rig in Alaska because of its drilling and lifting strength, the rig is expected to drill wells more than four miles long.[17]

"There's never been any wells like this (in the inlet), the long horizontal well with the huge multistage frack," Johnson said. The work won't pollute the inlet as has been the case in other states, he said. "The fracks don't go more than a few hundred feet from the well, and we're 7,000 feet deep under the seafloor."

In 2012, the U.S. Geological Survey found that technological advances in fracking could yield as much as 80 trillion cubic feet of natural gas and as much as 500 million barrels of recoverable natural gas liquids from Alaska's North Slope shale.[18] According to the Alaska Oil and Gas Conservation Commission, approximately 25 percent of the wells in Alaska are hydraulically fractured ("fracked").

In the *Alaska Dispatch News* article announcing BlueCrest's plans, Cathy Foerster, chair of the Alaska Oil and Gas Conservation Commission, described Alaska's fracking regulations as "very stringent."

"If BlueCrest makes an application that doesn't comply, we won't grant it," she said. "They'll either do it right, or they won't do it at all."

What does "do it right" mean in an earthquake-sensitive area like Cook Inlet, and with scientists exploring a link between quakes and fracking and the injection of fracking wastewater into underground disposal wells?[19] That question is not addressed in the ADN article. But Alaska is ranked by the USGS as the state with the most quakes—more than 50 percent of earthquakes nationwide. Between 1974 and 2003, 12,053 earthquakes of magnitude 3.5 or greater were reported in the state.[20] The 9.2 Good Friday earthquake of March 28, 1964, centered in the Prince William Sound area, was felt strongly in Cook Inlet and is one of the strongest ever recorded. U.S. Geological Survey scientist Peter Haeussler, who used seismic data collected by the oil and gas industry to illustrate in 2000 the web of faults beneath the inlet, warned of the potential of magnitude 6.0 and greater earthquakes and urged oil and gas companies to assess the vulnerability of their pipelines.[21] On the morning of Jan. 24, 2016, residents in the Cook Inlet area received a strong reminder of their area's vulnerability when they were shaken by a 7.1 earthquake.

Another factor in the "do it right" equation is the federal government's concern about the potential effect of oil and gas activity on the marine mammal population of the inlet. A March 28, 2016, letter from the U.S. Marine Mammal Commission to the National Marine Fisheries Service, which deals directly with the oil and gas industry, focuses first on beluga whales because of declining sightings in the area.[22] "There are a number of human and natural forces that could be contributing to the continued demise of beluga whales (in Cook Inlet); we just aren't sure what the circumstances are," Vicki Cornish, an energy policy analyst and marine biologist with the Marin Mammal Commission is quoted as saying, in a *Peninsula Clarion* article on April 19. But other species, including gray and minke whales and Dall's porpoises, are also of concern. "Each animal has value in the environment, and it's hard for us as people to understand that. All of them are connected, all of them have a role to play," Cornish says in the *Clarion* article.[23]

. .

WHATEVER BLUECREST'S plans may be, as of April 2, the date of the *Alaska Dispatch News* article, the company had yet to make those plans formal, according to Jody Colombie, special assistant with the Alaska Oil and Gas Conservation Commission. "We have no way of knowing what operators' plans are until they submit permit requests, and we have not received any permit requests for hydraulic fracturing in Cook Inlet," she wrote in our email exchange. When I asked if that meant no requests from BlueCrest, Jody said, "That is correct. They have not submitted a plan yet."

Concerns about BlueCrest's plans have led to the scheduling of a public forum to be held in Homer in mid-May, with representatives from BlueCrest, the Alaska Oil and Gas Conservation Commission and Cook Inletkeeper scheduled to be present.

FROM ALL APPEARANCES Hilcorp's interest in the Kenai Peninsula-Cook Inlet area remains strong.

On March 18, 2016, the Alaska Department of Natural Resources, Division of Oil and Gas, published a notice of competitive oil and gas lease sales to be held May 2, 2016—an Alaska Peninsula area-wide sale encompassing about four million gross onshore acres and 1.75 million gross acres of offshore state waters,[24] and a Cook Inlet area-wide lease sale encompassing about 4.2 million gross acres.[25]

"Right now, pretty exciting times—a lot of properties are probably going to be available for sale," Chad Helgeson, Hilcorp's Kenai area operations manager, reportedly said at a meeting in January of the Industry Outlook Forum. "What are we going to buy next? I have no idea."[26]

. .

JIM AND MARY Arndt aren't worried about Hilcorp's next purchase. They're worried about what's happening now, outside their home, and where it will lead if the seismic results are what Hilcorp hopes. In a conversation with a company spokesperson, Jim was told "they're planning on fracking everything. That doesn't give me a warm fuzzy feeling."

The seismic crew "seems to be pretty conscientious" when it comes to keeping the Arndts apprised of what the work involves, Jim said, but "the writing's on the wall. At some point down the line, they have the potential to come in and set up next to my house."

"It seems like the state is having their cake and eating it, too," Jim said. "We're not able to do anything that's not ag-related, but they (Hilcorp) can have their industrial complex? . . . Have you seen the statute regarding mineral rights? You cannot deny them 'reasonable access' to any of the resources at all. They can come and go at any time. They're supposed to let you know when, but it could be at any time."

Chapter Notes

Introduction

1. Election 2008, Transcript: The Vice-Presidential Debate, *The New York Times, May 23, 2012.*

2. "Governor charms Homer, and vice versa, it appears," by McKibben Jackinsky and Ben Stuart, *Homer News*, July 3, 2008.

3. "We know Sarah Palin," editorial, *Mat-Su Valley Frontiersman,* Aug. 30, 2008.

4. "Hardball with Chris Matthews," Friday, Oct. 10, 2008, http://www.nbcnews.com/id/27164004/ns/msnbc-hardball_with_chris_matthews/t/hardball-chris-matthews-friday-october/#.VyEqfnrGD-U.

5. "50th Anniversary Edition," *Homer News,* Jan. 1, 2015.

6. "Looking good for gas / Armstrong 'cautiously optimistic' about peninsula well, meeting with ENSTAR," by Kay Cashman, *Petroleum News*, Oct. 19, 2008.

7. "Producers 2013: Armstrong: southernmost producer / The Denver independent and its four partners are working to increase gas production at their signature Cook Inlet development," by Eric Lidji, *Petroleum News,* Nov. 17, 2013.

8. "Homer part of long-term gas plan, says ENSTAR," by Sean Pearson, *Homer Tribune*, Oct. 7, 2009.

9. "Southcentral prepared for emergency power needs," by Tim Bradner, *Alaska Journal of Commerce,* reprinted in the *Homer News*, Sept. 9, 2009.

10. "Cook Inlet Gas Study—2012 Update," prepared by Peter J. Stokes, PE, Petrotechnical Resource of Alaska, for ENSTAR Natural Gas Co., Chugach and Municipal Light and Power, October 2012.

Emil

1. "Homesteading and The Homestead Act in Alaska," http://www.alaskacenters.gov/homestead.cfm.

2. "History of Homesteading in Alaska," www.blm.gov/ak/.

3. The Alaska Volcano Observatory (AVO) reports lava flows were observed from Mount Augustine as early as December 1812. In 1883, according to observations noted in the Alaska Commercial Company's books from its location in the village now known as Nanwalek, the volcano erupted with such force that it spewed black clouds and ash into the sky, set off a thunderstorm and generated a tidal wave that raised sea level 20 feet. In 1944, three years before the Bartolowitses arrived in Alaska, formation of a small lava dome formed in Augustine's crater. After a period of relative quiet, in 1963 the volcano awoke violently with a plume of ash rising 3,000 feet and a string of eruptions that continued for the next 10 months. The restless volcano's most recent activity began with microearthquakes of one to two a day in May 2005. By December of that year, the quakes had increased to 15 a day. That progressed to what the AVO describes as an "explosive phase" from January 11–28, 2006. The event moved into its final phase in March 2006. Currently, Augustine is identified by the volcano observatory as "green," meaning it is in a noneruptive state. (avo.alaska.edu.)

4. Bald eagles are protected by the Bald and Golden Eagle Protection Act and the Migratory Bird Treaty Act. The former prohibits the "taking" of bald eagles, which includes their parts, nests or eggs. It provides penalties, both civil and criminal, for persons who "take, possess, sell, purchase, barter, offer to sell, purchase or barter, transport, export or import, at any time or any manner, any bald eagle . . . alive or dead, or any part, nest or egg thereof." The Act defines "take" as meaning "pursue, shoot, shoot at, poison, wound, kill, capture, trap, collect, molest or disturb." The definition of "disturb" covered by the Act is to agitate or bother an eagle so as to cause injury, a decrease in productivity or nest abandonment.

    Under the Migratory Bird Treaty Act the taking of any migratory bird or any part, nest or egg, except as permitted by regulation, is prohibited.

    Although not listed on the endangered species list for Alaska, the delisting of bald eagles elsewhere resulted in creation of the U.S. Fish and Wildlife Service's National Bald Eagle Management Guidelines of 2007. It includes specific guidelines for eight types of activity, including oil and natural gas drilling and refining and associated activities. Under that category, if there is no similar activity within a mile of the nest, the recommended distance from a visible nest is 660 feet, with landscape buffers recommended. If there is similar activity within a mile and the activity is visible from the nest, the recommended distance remains 660 feet "or as close as existing tolerated activity of similar scope," with landscape buffers recommended. If the activity is not visible from the nest, the recommended distance remains 660 feet if there is no similar activity within a mile of the nest, but is decreased to 330 feet if there is similar activity within a mile. (National Bald Eagle Management Guidelines, U.S. Fish and Wildlife Service, May 2007.)

ROBERT AND STACY JO

1. Cook Inletkeeper, Inletkeeper.org.

2. In 1976, with the 800-mile trans-Alaska pipeline under construction to deliver oil from the North Slope to tankers in Valdez, Alaska voters approved a constitutional amendment to create the Permanent Fund, an investment of the state's oil money set aside for the future. Four years later, Permanent Fund Dividend legislation was enacted that created a dividend fund and would give every adult Alaska resident $50 for every year of residency since Alaska had become a state in 1959.

    Payment of the dividends was challenged by Ron and Penny Zobel of Anchorage, who argued that the residency requirement discriminated against newer residents. The case went all the way to the U.S. Supreme Court. With the possibility that the court case would bring an end to the dividend plan, then-Gov. Jay Hammond devised a plan for equal payments to all residents with at least six months residence, with the first payment set at $1,000. On June 14, 1982, the court struck down the 1980 plan. Within hours, Hammond signed his plan into effect and the first checks were printed and mailed.

    In 1988, the state's Permanent Fund Dividend Division was created, and the following year the Legislature changed the residency period from six to 24 months. In 1990, the Superior Court cut that requirement in half, backing it up to 12 months.

    Since those first $1,000 checks were paid to 469,741 applicants in 1982, the annual payments have ranged from a low of $331 paid to 481,349 applicants in 1984 to a high of $2,072 and a database listing 664,163 applicants in 2015.

KATIE

1. "Spill: The wreck of the Exxon Valdez," final report, Alaska Oil Spill Commission, February 1990, state of Alaska.

2. "Buccaneer selling out / Bankruptcy independent seeks court approval for sale to lender AIX Energy," by Eric Lidji, *Petroleum News*, week of Nov. 2, 2014.

3. "CIRI settles long-running dispute over Kenai Loop gas fees," by Elwood Brehmer, *Alaska Journal of Commerce*, Jan. 28, 2015.

4. "Alaskan geologists warn of massive quake potential," *Alexander's Oil and Gas Connections*, Sept. 29, 2000.

5. "Application for the Incidental Harassment Authorization for the Taking of Non-listed Marine Mammals in Conjunction with the BlueCrest Alaska Operating LLC Activities at Cosmopolitan State Unit Alaska, 2014," prepared for BlueCrest Alaska Operating LLC by Owl Ridge Natural Resource Consultants, July 2014.

6. "Petition for Incidental Take Regulations for Seismic Program Cook Inlet, Alaska in 2015-2020," prepared for Apache Alaska Corp. by ASRC Energy Services, July 2014.

7. "High tides, wind stir up clams," by McKibben Jackinsky, *Homer News*, Nov. 24, 2010.

8. "Cause of razor clams' demise no mystery; it's overharvest," by Frank Mullen, *Homer News*, March 12, 2015.

9. "Species Profiles: Life Histories and Environmental Requirements of Coastal Fishes and Invertebrates (Pacific Northwest), Pacific Razor Clams," Biological Report 82(11.89) RF EL-82-4, January 1989, by Dennis R. Lassuy and Douglas Simons, performed for Coastal Ecology Group, U.S. Army Corps of Engineers and U.S. Department of the Interior, Fish and Wildlife Service.

10. Online at http://www.st-andrews.ac.uk/news/archive/2013/title,228355,en.php: "Scientists discover noise pollution effects on shellfish," Oct. 3, 2013.

11. "Governor vetoes $138 million; longevity bonus program discontinued / Murkowski slashes state budget," by Mike Chambers, Associated Press, published in the *Peninsula Clarion*, June 13, 2003; Longevity letter by Gov. Frank Murkowski, June 11, 2003.

12. http://coast.noaa.gov/czm/act/

13. "Coastal management initiative panned, praised in first hearing," by Jerzy Shedlock, *Peninsula Clarion*, reprinted in the *Homer* News, July 4, 2012.

14. "Battle lines drawn over coastal zone management initiative," by Becky Bohrer, Associated Press, in the *Alaska Dispatch News*, Aug. 12, 2012.

15. http://www.elections.alaska.gov/results/12PRIM/data/results.htm.

16. "Senator Micciche Earns the Most, Costs the Least," by Catie Quinn, KSRM Radio Group, March 18, 2015.

17. "Why can Alaska lawmakers vote with conflicts of interest?" by Laurel Andrews, *Alaska Dispatch News*, April 4, 2013.

18. "Micciche unfazed by criticism," by Brian Smith, *Peninsula Clarion*, April 22, 2013.

19. "'Vote Yes' rally draws big show of support," by McKibben Jackinsky, *Homer News*, June 18.

20. http://www.elections.alaska.gov/results/14PRIM/data/results.htm.

Darwin and Kaye

1. *Agrafena's Children: The Old Families of Ninilchik, Alaska,* first edition, edited by Wayne Leman, Agrafena Press, Hardin, Montana, 1993.

2. "Explorers 2015: Hilcorp exploring at Two Cook Inlet units," by Eric Lidji, *Petroleum News*, June 7, 2015.

3. According to Sperling's Best Places, a database that compares food, housing, utilities, transportation, and health costs to determine the cost of living, Ninilchik ranks 18.4 percent higher than the national average. Other Kenai Peninsula communities, including Anchor Point at 19.2 percent; Soldotna at 26.9 percent and Homer at 29.9 percent, also exceed the national average. Sperling's rates Anchorage even higher, at 43.3 percent.

    "Moving to Alaska," on alaska.net, says that living in Alaska is "affordable and significantly less expensive than San Francisco, Honolulu, Manhattan and a handful of other U.S. cities." Sperling's rates San Francisco at 142.6 percent and Honolulu at 99.2 percent above the national average (Manhattan was not listed by Sperling's).

4. Oil and Gas unit Fact Sheet, Deep Creek Unit, Alaska Division of
    Oil and Gas; "Explorers 2015: Hilcorp exploring at two Cook Inlet
    units," by Eric Lidji, *Petroleum News*, June 7, 2015.

5. Signed by President Abraham Lincoln in 1862, 15 years after the
    ancestors of Martha Kelly Jensen, the woman Darwin fondly de-
    scribed as his "Alaska mother," arrived in Ninilchik and five years
    before the United States purchased Alaska from Russia, the
    Homestead Act made it possible for U.S. citizens who were the head
    of a household or a minimum of 21 years of age to claim a 160-acre
    parcel of land in the public domain. The homesteader was required
    to reside on the land, build a home, make improvements and farm
    for five years. At the end of five years, if all conditions had been
    met, he or she was given full ownership of the acreage.
    (*The Homestead Act of 1862*) The Act was repealed in 1976, but
    provisions allowed homesteading to continue in Alaska for another
    decade. Overall in the United States, the Act transferred 270,000
    million acres, 10 percent of the country, to private ownership.
    ("About the Homestead Act—Homestead National Monument
    of America, U.S. National Park Service," www.nps.gov.) Within
    Alaska, 3,277 homesteads were conveyed, for a total of 360,000
    acres, less than 1 percent of the total land in the state.
    ("Homesteading and The Homestead Act in Alaska,"
    Alaska Public Lands Information Center,
    http://www.alaskacenters.gov/homestead.cfm.

6. http://education.nationalgeographic.org/encyclopedia/tidal-energy.
    Also, "Tidal giants—the world's five biggest tidal power plants,"
    power-technology.com, April 11, 2014.

7. National Ocean Service, National Oceanic and Atmospheric
    Administration, oceanservice.noaa.gov/facts/highesttide.html.

8. "Understanding Alaska's Revenue," www.alaskabudget.com.

9. "Gov. Walker: Conversations with Alaskans aim to build a
    sustainable future," *Alaska Dispatch News*, May 31, 2015.

10. "Governor releases 2017 budget legislation," State of Alaska
    Gov. Bill Walker press release, Dec. 15, 2015.

11. "After a year in office, Alaska Gov. Walker reflects, looks ahead,"
    by Becky Bohrer, Associated Press, Dec. 28, 2015.

12. "Bold Steps, New Beginnings: Pulling Together for Alaska,"
    State of the State, Gov. Bill Walker, Jan. 21, 2016, gov.alaska.gov.

Debbie

1. "Apache pauses its Ninilchik work," by McKibben Jackinsky, *Homer News*, September 27, 2012.

2. "Apache gets permit, but not one needed to restart Ninilchik work," by McKibben Jackinsky, *Homer News*, February 26, 2013.

3. "The Economic Impact of the Seafood Industry in Southcentral Alaska," by the McDowell Group for the Alaska Salmon Alliance, June 20, 2015.

4. Neal Fried, economist, state of Alaska Labor and Workforce Development.

5. "Apache oil company to shut down Alaska operations," by Alex DeMarban, *Alaska Dispatch News*, March 3, 2016.

My Roots II—Homecoming

1. "Motherland of the Spirit" is reproduced in *Any Tonnage, Any Ocean: Conversations with a resolute Alaskan / Capt. Walter Jackinsky Jr. of Ninilchik, 34-year veteran of the Alaska Marine Highway System*, Hardscratch Press, 2004, p. 198.

Lara

1. *Arctic Dance: The Mardy Murie Story*, page 86, by Charles Craighead and Bonnie Kreps, Graphic Arts Center Publishing, Portland, Ore., 2002.

2. *Midnight Wilderness: Journeys in Alaska's Arctic National Wildlife Refuge*, page 170, by Debbie S. Miller, Sierra Club Books, San Francisco, Calif., 1990.

3. *Midnight Wilderness*, page 62, and U.S. Fish and Wildlife Service's "Arctic National Wildlife Refuge" website, http://www.fws.gov/refuge/arctic/.

4. U.S. Energy Information Administration, eia.gov/tools/faqs/faq.cfm?id=33&t=6

5. "Arctic National Wildlife Refuge, 1002 Area, Petroleum Assessment, 1998, Including Economic Analysis," prepared by the U.S. Geological Survey, 1998. The 1002 area gets its nickname from the section of ANILCA in which its resources and the impacts of oil and gas exploration, development and production are addressed.

6. "Undiscovered oil resources in the Federal portion of the 1002 Area of the Arctic National Wildlife Refuge: An economic update," by E.D. Attanasi for the U.S. Department of the Interior and the U.S. Geological Survey, 2005.

7. ANWR Resource Estimates, from the Arctic Power website, http://www.anwr.org/features/pdfs/ANWR_estimates.pdf.

8. "ANWR: Producing American Energy and Creating American Jobs," naturalresources.house.gov/anwr/

9. https://www.whitehouse.gov/the-press-office/2016/02/22/ remarks-president-national-governors-association-reception.

10. "Governor pays contractor $12,500/month to advocate ANWR development," by Austin Baird, political, rural reporter, KTUU. http://www.ktuu.com/news/news/governor-pays-contractor-12500month-to-advocate-anwr-development/38181732.

11. *Crude Dreams: A Personal History of Oil and Gas Politics in Alaska*, page 130, by Jack Roderick, Epicenter Press, Fairbanks/Seattle, 1997.

12. *Midnight Wilderness*, page 62.

13. "A Historical Perspective of Oil and Gas Exploration and Development Activities on the Kenai National Wildlife Refuge," by Jim Frates, March 24, 1999.

14. "A Brief History of Oil Drilling in the Kenai National Wildlife Refuge," by Shana Loshbaugh, for an Alaska History class at the University of Alaska Fairbanks, NOR 664, Dec. 15, 2005.

15. http://www.lb7.uscourts.gov/documents/13-21421.pdf.

16. "National Wildlife Refuges—Opportunities to Improve the Management and Oversight of Oil and Gas Activities on Federal Lands, United States General Accounting Office," Report to Congressional Requesters, August 2003. http://www.gao.gov/products/GAO-03-517.

17. Alaska National Interest Lands Conservation Act, Public Law 96-487, Dec. 2, 1980, Section 1003, Prohibition on Development. http://www.nps.gov/legal/parklaws/Supp_V/laws1-volume1-anilca.pdf.

18. "Alaska congressional delegation blasts Obama's wilderness plan," staff report, *Fairbanks Daily News-Miner*, Jan. 25, 2015. http://www.newsminer.com/news/local_news/alaska-congressional-delegation-blasts-obama-s-wilderness-plan/article_13cded34-a4bd-11e4-96c7-4fe50970fbff.html.

19. adfg.alaska.gov.

20. "Court victory for wilderness in the Arctic Refuge," press release issued by the Alaska Wilderness League, Center for Biological Diversity, Defenders of Wildlife, Gwich'in Steering Committee, Northern Alaska Environmental Center, Resisting Environmental Destruction on Indigenous Lands, Sierra Club and The Wilderness Society, July 21, 2015. http://www.biologicaldiversity.org/news/ press_releases/2015/arctic-national-wildlife-refuge-07-21-2015.html.

21. Gwich'in Niintsyaa 2012, Resolution to Protect the Birthplace and Nursery Grounds of the Porcupine Caribou Herd. http://ourarcticrefuge.org/wp-content/uploads/2012/11/GG-Resol-2012.pdf.

MIKE AND JOANN

1. "Released to Reside Forever in the Colonies: Founding of a Russian-American Company Retirement Settlement at Ninilchik," by Katherine L. Arndt, presented at "The Anthropology of Cook Inlet," the 20th annual meeting of the Alaska Anthropological Association, Anchorage, Alaska, April 10, 1993.

2. *Agrafena's Children: The Old Families of Ninilchik, Alaska*, edited by Wayne Leman, Agrafena Press, Hardin, Montana, first edition, 1993.

ROBERT AND KATE

1. *In the Wake of the Exxon Valdez: The devastating impact of the Alaska oil spill*, by Art Davidson, Chapter 1, "Passages Through the Sound," Sierra Club Books, San Francisco, 1990. A barrel of oil equals 42 gallons; the exact spill figures, according to EVOSTC, the Exxon Valdez Oil Spill Trustee Council, were 1,264,155 barrels, or 53,094,510 gallons.
http://www.evostc.state.ak.us/index.cfm?FA=facts.QA

2. "Spill: The wreck of the Exxon Valdez," final report, Alaska Oil Spill Commission, February 1990, state of Alaska.

3. Navigation Center, U.S. Coast Guard, Department of Homeland Security, navcen.uscg.gov/.
Sound Truth and Corporate Myth$: The Legacy of the Exxon Valdez Oil Spill, by Riki Ott, Dragonfly Sisters Press, Cordova, Alaska, 2005. Ott, a fisherman in Cordova, holds several advanced degrees including a Ph.D. from the University of Washington School

of Fisheries with an emphasis on effects of heavy metals on benthic invertebrates.

4. *White Silk and Black Tar, a Journal of the Alaska Oil Spill*, by Page Spencer, page 27, Bergamot Books, Minneapolis, 1990.

5. *White Silk and Black Tar.*

6. *In the Wake of the Exxon Valdez.*

7. evostc.state.ak.us.

8. "Restoring Alaska, Legacy of an Oil Spill." Vol. 26, No. 1, Quarterly 1999, *Alaska Geographic*, Anchorage, Alaska.

9. "Residents still carry spill's hurt," by McKibben Jackinsky, *Homer News*, March 19, 2014.

10. "Jackup rig Endeavour arrives in Kachemak Bay," staff report, *Homer News*, Aug. 24, 2012; "Endeavour's delay fattens wallets of workers, city," by Michael Armstrong, *Homer News*, Dec. 5, 2012.

11. "Endeavour drilling contractor fired for 'non-performance,'" by Brian Smith, Morris News Service-Alaska, Dec. 17, 2012.

12. "Buccaneer files counterclaim in suit," by Michael Armstrong, *Homer News*, March 20, 2013.

13. "Endeavour leaving Cook Inlet," by Michael Armstrong, *Homer News*, Nov. 19, 2014.

14. "Restoring Alaska, Legacy of an Oil Spill."

15. "Fireweed students give fresh eyes to a 25-year-old disaster," by McKibben Jackinsky, *Homer News*, Jan. 29, 2014.

16. "Residents still carry spill's hurt."

17. Waterkeeper Alliance, waterkeeper.org.

18. Cook Inletkeeper, inletkeeper.org.

19. "Cook Inlet, Alaska: Oceanographic and Ice Conditions and NOAA's 18-Year Oil Spill Response History, 1984-2001, HAZMAT Report 2003-001," by John Whitney, NOAA scientific coordinator for Alaska, October 2002. http://docs.lib.noaa.gov/noaa_documents/NOS/HMRA/HAZMAT_report_2003-01.pdf.

20. "Massive ice floes in Cook Inlet break mooring lines of tanker," by Richard O. Aichele, professionalmariner.com, March 28, 2007. http://www.professionalmariner.com/March-2007/Massive-ice-floes-in-Cook-Inlet-break-mooring-lines-of-tanker/.

21. "State Reaches Settlement on Seabulk Pride Spill and Grounding," press release from the State of Alaska Department of Law, July 1, 2010.

22. "Oil terminal sits in harm's way," by Tom Kizzia, *Alaska Daily News*, Jan. 30, 2009. http://www.adn.com/print/article/20090130/oil-terminal-sits-harms-way.

23. "Mud flows in Drift River / Oil terminal status uncertain," by Richard Mauer, *Anchorage Daily News*, March 26, 2009, http://www.adn.com/article/20090326/mud-flows-drift-river-oil-terminal-status-uncertain; "Volcano forces Chevron to suspend Inlet production," by Kyle Hopkins, *Anchorage Daily News*, April 5, 2009, http://www.adn.com/article/20090405/volcano-forces-chevron-suspend-inlet-oil-production; "Tanker removed oil from Drift River Terminal," Alaska Public Media, May 1, 2009; "Hilcorp Alaska looks to resume oil storage near volcano," by Wesley Loy, *Petroleum News*, June 9, 2012, printed in the *Alaska Daily News*; June 9, 2012, http://www.adn.com/article/20120609/hilcorp-alaska-looks-resume-oil-storage-near-volcano.

MIKE

1. "Native American Uses of Asphaltum," Pacific Coastal Marine Sciences Center, U.S. Geological Survey, walrus.wr.usages.gov/seeps/native_uses.html.

2. "Oil and Gas Production History in California," California Department of Conservation, Division of Oil, Gas and Geothermal Resources, ftp://ftp.consrv.ca.gov/pub/oil/history/History_of_Calif.pdf.

3. "A Brief History of Oil and Gas Exploration in the Southern San Joaquin Valley of California," by Kenneth I. Takahashi and Donald L. Gautier, chapter 3 of "Petroleum Systems and Geologic Assessment of Oil and Gas in the San Joaquin Basin Province, California," U.S. Geological Survey Professional Paper 1713, edited by Allegra Hosford Scheirer, 2007.

4. "A Brief History of Oil and Gas Exploration in the Southern San Joaquin Valley of California."

5. www.bakersfieldchamber.org.

6. http://pubs.usgs.gov/fs/2012/3050/fs2012-3050.pdf.

7. "A Brief History of Oil and Gas Exploration in the Southern San Joaquin Valley of California."

8. The Ocean Portal Team, Smithsonian Institution National Museum of Natural History, http://ocean.si.edu/gulf-oil-spill.

9. "The Oil and Gas Industry in California, Its Economic Contribution and Workforce in 2013," Christine Cooper, Ph.D., and Shannon Sedgwick, Institute for Applied Economics, Los Angeles County Economic Development Corp., Los Angeles, California, June 2015. http://www.wspa.org/sites/default/files/uploads/2015%20O&G_Industry%20and%20Workforce.pdf.

10. "Drilling in California: Who's at risk?" by Tanja Srebotnjak and Miriam Rotkin-Ellman for the Natural Resource Defense Council, October 2014. http://www.nrdc.org/health/files/california-fracking-risks-report.pdf.

11. "Californians at Risk: An Analysis of Health Threats from Oil and Gas Pollution in Two Communities," prepared by Earthworks, in partnership with Clean Water Fund.

12. "Plunging oil prices choke off boom in Bakersfield," by Tiffany Hsu, Los Angeles Times, Jan. 29, 2015. http://www.latimes.com/business/la-fi-oil-bakersfield-20150129-story.html.

13. "Hundreds of illicit oil wastewater pits found in Kern County," by Julie Cart, Los Angeles Times, Feb. 26, 2015. http://www.latimes.com/local/lanow/la-me-ln-pits-oil-wastewater-20150226-story.html.

14. Alaska Oil and Gas Association. www.aoga.org.

15. "Opposition to oil permit mounts," Homer Weekly News, May 23, 1974.

16. "Kachemak fishermen file suit," Homer Weekly News, Dec. 12, 1974.

17. "Oil? Well, well, well . . . ," Homer Weekly News, Jan. 30, 1975.

18. "Judge puts 'laches' on fisherman's suit," Homer Weekly News, May 22, 1975.

19. "Agreement reached on Kachemak Bay leases," Homer News, Jan. 13, 1977.

20. "Shell leases repurchased," by Steve Cline, Homer News, July 14, 1977.

21. "Little Homer museum saved spill for history," by Alan Boraas, Alaska Dispatch News, June 11, 2010.

22. "Oil Spill's Impact Not Dying Down / Exxon Official Calls Exhibit 'Misleading,'" by Lisa T Stemle, special to the *Sun-Sentinel*, May 22, 1992.

23. www.nature.org.

GEE

1. Sonotube.com.

2. "Homeowners seek solution to erosion," by McKibben Jackinsky, *Homer News*, Nov. 6, 2014.

3. "Homeowners seek solution to erosion."

4. "HEA work sparks erosion concerns," by Michael Armstrong, *Homer News*, Aug. 10, 2006.

5. "City revokes HEA construction permit / HEA stops all work in bluff subdivision," by Michael Armstrong, *Homer News*, Sept. 1, 2006.

6. "Homeowners seek solution to erosion."

7. "Homer: City on the move," by McKibben Jackinsky, *Homer News*, Sept. 13, 2014.

8. "Homer: City on the move."

9. "Ditchwork may give some relief to erosion-plagued subdivision," by McKibben Jackinsky, *Homer News*, Dec. 4, 2014.

10. "Ditchwork may give some relief to erosion-plagued subdivision."

JOHNNY AND VIOLA

1. "BlueCrest plans Cook Inlet oil development," by Alan Bailey, *Alaska Dispatch News*, Sept. 21, 2014, http://www.adn.com/article/20140921/bluecrest-plans-cook-inlet-oil-development. "Buccaneer Energy—Cosmopolitan #1 Well Spuds," published by *Alaska Business Monthly*, May 16, 2013, http://www.akbizmag.com/Alaska-Business-Monthly/May-2013/Buccaneer-Energy-Cosmopolitan-1-Well-Spuds/; "BlueCrest Energy Activity Update," by J. Benjamin Johnson, as presented to CIRCAC Board of Directors meeting, Kodiak, Sept. 12, 2014, http://www.circac.org/wp-content/uploads/BlueCrest-Energy-9-12-2014.pdf.

2. "BlueCrest Energy Activity Update."

3. "Project Fact Sheet," BlueCrest Energy, June 5, 2015; "BlueCrest Energy Activity Update," by J. Benjamin Johnson to CIRCAC Board of Directors meeting in Kodiak, Sept. 12, 2014.

4. "BlueCrest energy expands operations in Cook Inlet," by Jenny Neyman, KDLL-Kenai, July 20, 2015.

5. "BlueCrest nearing end of construction phase," by Jenny Neyman, *Redoubt Reporter*, Soldotna, Alaska, Jan. 27, 2016.

6. "AIDEA Board Approves Cook Inlet Oil Field Project: Financing for BlueCrest Energy, Inc. On-Shore Drilling Rig," *Alaska Business Monthly*, Aug. 14, 2015.

HERFF, JOLAYNE, KEN

1. "Eastern Cook Inlet beaches closed to clamming," Alaska Department of Fish and Game news release, Feb. 24, 2015.

2. "Project Fact Sheet," BlueCrest Energy, June 5, 2015; "BlueCrest Energy Activity Update," by J. Benjamin Johnson to CIRCAC Board of Directors meeting in Kodiak, Sept. 12, 2014.

3. Noise Comparisons—Purdue University, https://www.chem.purdue.edu/chemsafety/Training/PPETrain/dblevels.htm.

4. "BlueCrest's construction phase nearing completion," by Jenny Neyman, *Redoubt Reporter*, Soldotna, Alaska, Jan. 27, 2016.

5. "Area still feeling impact of last year's floods," by Carly Bossert, *Homer News*, Oct. 30, 2003; "New bridge coming for Stariski," by McKibben Jackinsky, *Homer News*, April 26, 2007; "Rain washes out Homer again," by Chris Bernard, *Homer News*, Nov. 28, 2002; "State unveils plan for shoring up bluff to protect highway," by McKibben Jackinsky, *Homer News*, Oct. 31, 2012.

6. "BlueCrest Energy Activity Update."

7. Homer City Council meeting minutes, Oct. 13, 2008.

8. "Homer gas project nixed by governor," by Michael Armstrong, *Homer News*, June 9, 2010; "Guests, nonagenarian, gas line dominate meeting," by McKibben Jackinsky, *Homer News*, Aug. 25, 2010.

9. Ordinance 13-03(S)(2), Homer City Council, Feb. 25, 2013.

10. "Oil and Gas at Your Door? A Landowner's Guide for Oil and Gas Development," Oil and Gas Accountability Project, 2004.

11. "Facts About Benzene," Centers for Disease Control and Prevention, www.bt.cdc.gov/agent/benzene/basics/facts.asp.

12. "Anchor Point takes steps to get natural gas flowing to residents," by McKibben Jackinsky, *Homer News*, July 14, 2011.

13. http://www.anchorpointchamber.org/index.php/history.

14. *Alaska's Kenai Peninsula: The Road We've Traveled*, published by the Kenai Peninsula Historical Society, 2002; *Snapshots at Statehood: A Focus on Communities that Became the Kenai Peninsula Borough*, published by the Kenai Peninsula Historical Society, 2009.

15. census.gov.

16. "Superfund" is the name of an environmental program that addresses abandoned hazardous waste sites. It also is the name of the fund established by the Comprehensive Environmental Response, Compensation and Liability Act of 1980, which allows the U.S. Environmental Protection Agency to clean up the sites and compels responsible parties to either clean up or reimburse for EPA-led cleanups. On the National Priorities List identifying hazardous waste sites in the country that are eligible for extensive, long-term cleanup under the Superfund program, New Jersey is tops with 114 sites. California is second with 97 and Pennsylvania third with 95. Alaska has six. www.epa.gov.

LEADERSHIP

1. Alaska's Oil and Gas Industry: A look at jobs and oil's influence on economy," by Neal Fried in the Alaska Department of Labor's *Alaska Economic Trends*, June 2013. http://labor.alaska.gov/trends/jun13.pdf

2. "Population Growth and Migration in Alaska," by Greg Williams in the Alaska Department of Labor's *Alaska Economic Trends*, October 1985. http://laborstats.alaska.gov/trends/oct85art1.pdf.

3. "540 say no to gas proposal," by McKibben Jackinsky, *Homer News*, Jan. 31, 2013.

4. City of Homer Resolution 15-047(S) (A), June 15, 2015, "Changing Dates for Payment of Assessments for the Homer Natural Gas Special Assessment District." http://www.cityofhomer-ak.gov/resolution/resolution-15-047sa-changing-dates-payment-assessments-homer-natural-gas-special.

5. "Council extends grace period for gas assessments," by Michael Armstrong, *Homer News*, Sept. 1, 2015.

6. McDowell Economic Impact Report, www.aoga.org.

7. "Governor releases amended endorsed budget," gov.alaska.gov/Walker.

8. ak-lng.com.

9. The AK LNG Project, Alaska Natural Gas Transportation Projects, Office of the Federal Coordinator, www.arcticgas.gov.

10. "TransCanada: Alaska LNG Project no longer 'commercially reasonable' for company," by Austin Baird, KTUU, Oct. 30, 2015. http://www.ktuu.com/news/news/transcanada-alaska-lng-project-no-longer-commercially-reasonable/36150680.

11. Statement by the President on the Keystone XL Pipeline, Nov. 6, 2015, https://www.whitehouse.gov/the-press-office/2015/11/06/statement-president-keystone-xl-pipeline.

12. "Signed gas agreements made public," Dec. 8, 2015, http://gov.alaska.gov/Walker/press-room/full-press-release.html?pr=7345.

13. "Nikiski residents move to make way for AK LNG," by Lacie Grosvold Leichliter, KTUU, March 3, 2015. http://www.ktuu.com/news/news/nikiski-residents-move-to-make-way-for-ak-lng/31600916.

14. "Persily hired by borough mayor as oil-gas adviser," by Rashah McChesney, Morris News Service, *Homer News*, March 11, 2015.

15. "Producers agree to $16.5B for PILT, AK LNG impact payments," by Elwood Brehmer, *Alaska Journal of Commerce*, Sept. 30, 2015. http://www.alaskajournal.com/2015-09-30/producers-agree-165b-pilt-ak-lng-impact-payments#.Vtgq95MrKb8.

16. "Administration rolls out fiscal plan," http://gov.alaska.gov/Walker/press-room/full-press-release.html?pr=7346.

17. "Order Granting Long-Term Multi-Contract Authorization to Export Liquefied Natural Gas By Vessel From the Proposed Alaska LNG Project In the Nikiski Area of the Kenai Peninsula, Alaska, To Free Trade Agreement Nations," DOE/FE Order No. 3554, Nov. 21, 2014, http://energy.gov/sites/prod/files/2014/11/f19/ord3554%20fta_0.pdf.

18. "Order Conditionally Granting Long-Term, Multi-Contract Authorization to Export Liquefied Natural Gas By Vessel From the

Proposed Alaska LNG Terminal in Nikiski, Alaska, to Non-Free Trade Agreement Nations," DOE/FE Order No. 3643, May 28, 2015.

19. http://www.crowley.com/News-and-Media/Press-Releases/Video-Crowley-Now-Authorized-to-Import-LNG-for-Distribution-in-Pacific-Northwest-and-Alaska.

20. "Crowley takes step to import Canadian LNG to Alaska," by Alex DeMarban, *Alaska Dispatch News*, Nov. 11, 2015.

21. "Endeavour leaving Cook Inlet," by Michael Armstrong, *Homer News*, Nov. 19, 2014.

Living With Less

1. "Global Climate Change, Vital Signs of the Planet," climate.nasa.gov.

2. "The Impact of Global Warming on Human Fatality Rates," *Scientific American*, June 17, 2009. http://www.scientificamerican. com/article/global-warming-and-health/.

3. "HEA ready to plug into hydro / Grant Lake project heading on to licensing," by Jenny Neyman, *Redoubt Reporter*, Feb. 25, 2015. http://peninsulaclarion.com/news/2015-04-04/ grant-lake-hydroelectric-project-open-to-comments.

4. "HEA ready to plug into hydro."

5. "Homer Electric Association Presents Case for Hydro Plant Near Moose Pass," *Seward City News*, Nov. 9, 2014. http://sewardcitynews.com/2014/11/homer-electric-association-presents-case-for-hydro-plant-near-moose-pass/.

6. "Grant Lake hydroelectric project open to comments," by Ben Boettger, *Peninsula Clarion*, April 4, 2015. http://peninsulaclarion. com/news/2015-04-04/grant-lake-hydroelectric-project-open-to-comments.

7. "Kodiak reaps benefits of renewable energy, with lessons for rural Alaska," by Michelle Theriault Boots, *Alaska Dispatch News*, Sept. 26, 2015. http://www.adn.com/article/20150926/ kodiak-reaps-benefits-renewable-energy-lessons-rural-alaska.

8. "TDX to Power St. Paul Island with 80 Percent Renewable Energy," by Kailee Wallis, *Alaska Business Monthly*, November 2015.

9. Federal Laboratory Consortium for Technology Transfer, www.federallabs.org.

10. In 1971, T. Theodore Fujita of the University of Chicago created the Fujita Tornado Damage Scale. The least severe by Fujita's ranking was an F0 with winds up to 73 miles per hour and capable of causing light damage. The most severe was an F5, with winds at 261–318 mph, powerful enough to cause "incredible damage." An Enhanced F Scale for Tornado Damage was created by a team of meteorologists and wind engineers and implemented in 2007. By that scale, the least severe is an EF0 that has three-second wind gusts of 65–85 mph and the most severe an EF5 with three-second wind gusts exceeding 200 mph. (Storm Prediction Center, National Oceanic and Atmospheric Administration, www.spc.noaa.gov.)

11. "What are the most commonly traded commodities?" InvestorGuide staff, www.investorguide.com.

12. "The World's Largest Companies 2015," by Liyan Chen, *Forbes* staff, forbes.com.

13. "Which companies in the Fortune 500 earned the most last year?" fortune.com.

14. "A Closer Look At The 5 Biggest Oil Companies," by Gaurav Agnihotri, oilprice.com, April 2015.

15. "Oil Demand Growing at Fastest Pace in Five Years, Says IEA," by Benoit Faucon, *Wall Street Journal*, Aug. 12, 2015. http://www.wsj.com/articles/oil-demand-growing-at-fastest-pace-in-five-years-says-iea-1439367413.

16. energy.gov.

17. "ASLC shifting 98% of the Center's heating needs from fossil fuel to ocean water as source heat," Alaska SeaLife Center website, April 22, 2016, www.alaskasealife.org/news_items/34. The CO_2 refrigerant heat pumps push seawater through a titanium-plate heat exchanger, the cooled seawater is returned to the ocean, and the converted, captured heat runs through a high pressure system that blends into the main building's heat loop. The installation is an example of how "our day to day work can contribute to the long term health and sustainability of the city of Seward, the state of Alaska and the global community," Darryl Schaefermeyer, the center's special projects director, said in the web posting.

My Roots IV—Writing

1. From *Seasonal Woman*, by Luci Tapahonso,
   Tooth of Time Books, ©1982 Luci Tapahonso.
   http://www.hanksville.org/storytellers/luci/poems/prayer.html.

2. "Growing with Alaska in mind," by Aryn Young, *Homer News*,
   Aug. 6, 2015. http://homernews.com/homer-news/local-news/
   2015-08-06/growing-with-alaska-in-mind.

Describing the Elephant

1. www.esrl.noaa.gov

2. climate.nasa.gov/effects/.

3. "Unocal, Marathon announce gas discovery at Ninilchik,"
   by Kristen Nelson, *Petroleum News Alaska*, Jan. 27, 2002.

4. *Early Warming, Crisis and Response in the Climate-Changed North*,
   by Nancy Lord, Counterpoint, Berkeley, Calif., 2011.

5. https://www.whitehouse.gov/sites/default/files/microsites/ostp/
   PCAST/pcast_energy_and_climate_3-22-13_final.pdf. Also, "Top
   U.S. Scientist: World Must Act Now to Reverse Climate Change,"
   by Craig Welch, *National Geographic*, Dec. 7, 2015. http://news.
   nationalgeographic.com/2015/12/151207-climate-change-
   holdren-white-house-science-paris/.

6. Widely quoted, the full text of Palin's speech can be found at
   buzzfeed.com.

7. *Fuel*, directed by Josh Tickell; produced by Greg Reitman, Dale
   Rosenbloom, Daniel Assael, Darius Fisher and Rebecca Harrell
   Tickell; 2008; winner of the 2008 Sundance Film Festival
   audience award.

8. *In the Wake of the Exxon Valdez, The Devastating Impact of the
   Alaska Oil Spill*, by Art Davidson, page 315, Sierra Club Books,
   San Francisco, Calif., 1990.

9. "Hilcorp pays $115,000 penalty for drilling violations,"
   by Wesley Loy, *Alaska Dispatch News*, June 14, 2013.

10. "No. 5 midsize company: Hilcorp," by Tayna Rutledge, *Houston
    Chronicle*, Nov. 5, 2015; "50 Best Workplaces for Diversity, 2015:
    27, Hilcorp," *Fortune*. fortune.com.

11. "How did three North Slope workers narrowly avert death?" by Alex DeMarban, *Alaska Dispatch News*, Dec. 23, 2015.

JIM

1. "BP to cut 13 percent of Alaska workforce as oil prices keep dropping," by Alex DeMarban, *Alaska Dispatch News*, Jan. 12, 2016.

2. "Gov. Walker says BP decision to cut rigs underscores need for fiscal solution to state's budget crisis," by Paula Dobbyn, KTUU, March 8, 2016.

3. "Apache oil company to shut down Alaska operations," by Alex DeMarban, *Alaska Dispatch News*, March 3, 2016.

4. "Alaska tax credit payouts to oil companies rival oil revenue," by Austin Baird, KTUU, March 27, 2016.

5. "Alaska lawmakers are looking at changes in oil and gas tax credits," by Becky Bohrer, Associated Press, printed in the *Alaska Dispatch News*, April 1, 2016.

6. "'No way' to avoid overtime lawmaking, senate president says," by Austin Baird, KTUU, April 16, 2016, www.ktuu.com/ news/news/no-way-to-avoid-overtime-lawmaking-senate-president-says/39069880.

7. "Governor: Expect extra time in Legislature," by James Brooks, *Juneau Empire*, April 17, 2016, juneauempire.com/state/ 2016-04-17/expect-extra-time-legislature.

8. "Legislature extends its session with only hints of progress on key issues," by Nathaniel Herz, *Alaska Dispatch News*, April 18, 2016, http://www.adn.com/article/20160417/alaska-legislature-extends-session-only-hints-progress-key-issues.

9. http://elections.alaska.gov/results/06GENR/data/results.htm.

10. Senate Journal, March 19, 2013. http://www.akleg.gov/basis/get_ single_journal.asp?session=28&date=20130319&beg_ page=0621&end_page=0639&chamber=S.

11. "Bill History/Action for the 28th Legislature," Senate Bill 21, http:// www.akleg.gov/basis/get_bill.asp?bill=SB%20%2021&session=28.

12. "Why can Alaska lawmakers vote with conflicts of interest?" by Laurel Andrews, *Alaska Dispatch News,* April 4, 2013, www.adn.com/print/article/20130404/why-can-alaska-lawmakers-vote-conflicts-interest.

13. "Climate change challenges subsistence hunters in Unalakleet," KTUU, March 30, 2016, http://www.ktuu.com/news/news/climate-change-challenges-subsistence-hunters-in-unalakleet/38784890.

14. "Range Expansion of Moose in Arctic Alaska Linked to Warming and Increased Shrub Habitat," by Ken D. Tape, David D. Gustine, Roger W. Ruess, Layne G. Adams, Jason A. Clark, April 13, 2016, http://journals.plos.org/plosone/article?id=10.1371/journal.pone.0152636.

15. "What Exxon knew about the Earth's melting Arctic," by Sara Jerving, Katie Jennings, Masako Melissa Hirsch and Susanne Rust, *Los Angeles Times,* Oct. 9, 2015.

16. "Big Oil braced for global warming while it fought regulations," by Amy Lieberman and Susanne Rust, *Los Angeles Times,* Dec. 31, 2015, graphics.latimes.com/oil-operations/#about. "Fighting regulations" has not been inexpensive, according to another group of researchers: "Oil Giants Spend $115 Million A Year To Oppose Climate Policy," by Casey Williams, Editorial Fellow, The Huffington Post, April 11, 2016, http://www.huffingtonpost.com/entry/oil-companies-climate-policy_us_570bb841e4b0142232496d97.

17. "Groundbreaking fracking effort, plus first new oil production in years, on tap in Cook Inlet," by Alex DeMarban, *Alaska Dispatch News,* April 2 2016, www.adn.com/print/article/20160402/groundbreaking-fracking-effort-plus-first-new-oil-production-years-tap-cook-inlet.

18. "Assessment of Potential Oil and Gas Resources in Source Rocks of the Alaska North Slope, 2012," National Oil and Gas Assessment Project, U.S. Department of Interior and U.S. Geological Survey, Fact Sheet 2012-3013, February 2012, http://pubs.usgs.gov/fs/2012/3013/pdf/fs2012-3013_2-28-2012.pdf.

19. "Introduction to Hydraulic Fracturing," U.S. Geological Survey, Feb. 25, 2015, http://www.usgs.gov/hydraulic_fracturing/.

20. "Top Earthquake States," USGS, http://earthquake.usgs.gov/earthquakes/states/top_states.php.

21. "Alaskan geologists warn of massive quake potential," Alexander's Oil and Gas Connections, Sept. 29, 2000.

22. Letter to Jolie Harrison, Chief, Permits and Conservation Division, Office of Protected Resources, National Marine Fisheries, from Rebecca J. Lent, Executive Director, Marine Mammal Commission, March 28, 2016, http://www.mmc.gov/wp-content/uploads/ 16-03-28-Harrison-Draft-EA-2016-oil-and-gas-IHAs-in-Cook-Inlet. pdf.

23. "Federal agency calls for comprehensive approach to beluga protection," by Elizabeth Earl, *Peninsula Clarion*, April 19, 2016, http://peninsulaclarion.com/news/2016-04-19/federal-agency-calls-for-comprehensive-approach-to-beluga-protection. One sentence in the *Clarion* article may be overstating the MMC's position (see previous note): "The commission asserts that the agency should not allow the permits because the cause of the decline of the Cook Inlet beluga population is unknown."

24. The Alaska Peninsula lease sale area is located west of Cook Inlet and includes tracts within the Aleutians East, Bristol Bay and Lake and Peninsula boroughs, as well as the Dillingham Census Area. Sale announcement and instructions to bidders, Alaska Peninsula Areawide 2016W Competitive Oil and Gas Lease Sale, http:// www.dog.dnr.alaska.gov/Leasing/Documents/SaleDocuments/ AKPeninsula/LatestSale/AP2016-SaleAnnouncement-BidderInstructions.pdf.

25. The Cook Inlet tracts are within the Municipality of Anchorage and the Matanuska-Susitna and Kenai Peninsula boroughs. The selected on- and off-shore areas in the Kenai Peninsula Borough stretch on the north from Turnagain Arm south to Anchor Point and Tuxedni Bay. They are bordered on the west by the Aleutian Range and on the east by the Chugach and Kenai mountains. Sale Announcement and Instructions To Bidders, Cook Inlet Areawide 2016W Competitive Oil and Gas Lease Sale, http://www.dog.dnr.alaska.gov/ Leasing/Documents/SaleDocuments/CookInlet/LatestSale/CI2016-SaleAnnouncement-BidderInstructions.pdf.

26. "Hilcorp still ready to buy assets as it looks to cost control," by Elizabeth Earl, *Peninsula Clarion*, printed in the *Alaska Journal of Commerce*, Feb. 3, 2016.

References & Recommended Reading

Many of the books, articles, agencies and websites consulted during research for this book appear in the Chapter Notes immediately preceding these References. Other sources consulted but not necessarily cited in individual chapters are listed here. As stressed in the final chapter, coverage of the oil and gas industry appears almost daily in newspapers and online postings, so even as these pages were readied for publication the list of recommended readings has continued to grow. (In our brave new e-world, website links are changed occasionally; we've found that in those cases a search for the name of the publication or agency in question will usually yield the new link.)

Books & Pamphlets

100 Ways To Save the World, by Johan Tell, Gold Street Press,
    San Francisco, 2008.

Arctic Dance: The Mardy Murie Story, by Charles Craighead and Bonnie
    Kreps, Graphic Arts Center Publishing, Portland, Oregon, 2002.

Agrafena's Children: The Old Families of Ninilchik, Alaska,
    edited by Wayne Leman, Agrafena Press, Hardin, Montana, 1993.

Crude Dreams, A Personal History of Oil and Politics in Alaska, by Jack
    Roderick, Epicenter Press, Fairbanks/Seattle, 1997.

*Darkened Waters, A Review of the History, Science, and Technology
    Associated with the Exxon Valdez Oil Spill and Cleanup*, by Nancy
    Lord, a publication of the Homer Society of Natural History/Pratt
    Museum to accompany the "Darkened Waters: Profile of an Oil Spill"
    exhibit, 1992.

Diapering the Devil: A Lesson for Oil Rich Nations, by Jay Hammond,
    Kachemak Resource Institute, Homer, Alaska, 2011.

Early Warning: Crisis and Response in the Climate-Changed North,
    by Nancy Lord, Counterpoint Press, Berkeley, Calif., 2011.

*A Field Philosopher's Guide to Fracking: How one Texas town stood up
    to Big Oil and Gas*, by Adam Briggle, Liveright Publishing,
    New York, 2015.

*In the Wake of the Exxon Valdez: The Devastating Impact of the Alaska
    Oil Spill,* by Art Davidson, Sierra Club Books, San Francisco, 1990.

Last Great Wilderness: The Campaign to Establish the Arctic National Wildlife Refuge, by Roger Kaye, University of Alaska Press, Fairbanks, 2006.

Midnight Wilderness: Journeys in Alaska's Arctic National Wildlife Refuge, by Debbie S. Miller, Sierra Club Books, San Francisco, 1990.

My Wilderness, by William O. Douglas, Comstock Publishing, 1989.

Not One Drop: Betrayal and Courage in the Wake of the Exxon Valdez Oil Spill, by Riki Ott, Chelsea Green Publishing Company, White River Junction, Vermont, 2008.

Oil and Honey: The Education of an Unlikely Activist, by Bill McKibben, St. Martin's Griffin, New York, 2013.

Pioneering Conservation in Alaska, by Ken Ross, University Press of Colorado, Boulder, 2006.

The Prize: The Epic Quest for Oil, Money and Power, by Daniel Yergin, Simon and Schuster, Great Britain, 2012.

Red: Passion and Patience in the Desert, by Terry Tempest Williams, Vintage Books, a division of Random House, New York, 2002.

Repeal the Giveaway, pamphlet from Vote YES on Proposition 1, Vic Fischer, chair, 2014.

Restoring Alaska: Legacy of an Oil Spill, Vol. 26, No. 1, Quarterly 1999, Alaska Geographic, Anchorage, 1999.

Sound Truth & Corporate Myth$: The Legacy of the Exxon Valdez Oil Spill, by Riki Ott, Dragonfly Sisters Press, Cordova, Alaska, 2005.

This Changes Everything, by Naomi Klein, Simon and Schuster, New York, 2014.

White Silk and Black Tar: A Journal of the Alaska Oil Spill, by Page Spencer, Bergamot Books, Minneapolis, 1990.

Videos, DVDs, CD-ROM

Alaska: Technology & Time, a film by Rick Wise, Non Fiction Television, July 25, 1979.

America's Wildest Refuge: Discovering the Arctic National Wildlife Refuge,

DVD, an Artery Industries Production in partnership with Alaska Geographic and the U.S. Fish and Wildlife Service in association with National Fish and Wildlife Foundation and Teya Technologies, 2010.

Deadly Neighbor: Living With Oil in Alaska, produced by Greenpeace, 1994.

Fuel, directed by Josh Tickell; produced by Greg Reitman, Dale Rosenbloom, Daniel Assael, Darius Fisher and Rebecca Harrell Tickell; 2008.

Kachemak Bay Ecological Characterization. CD-ROM, Kachemak Bay Research Reserve and National Oceanic and Atmospheric Administration, Coastal Services Center, 2001. NOAA/CSC/20017-CD, Charleston, S.C.

PERIODICALS, RADIO, TELEVISION

Note: In July 2014, the *Anchorage Daily News* became *Alaska Dispatch News*, still often referred to as *ADN*. The *Homer News* has been published under several names, beginning in 1950 under its present name; then beginning in 1955, *Kenai Peninsula Pioneer*; 1964, *Homer News*; 1973, *Homer Weekly News*; 1974-present, *Homer News*. Other Homer newspapers: the *Homer Homesteader*, 1944-49; *Homer Herald*, 1957-58; *Cook Inlet Courier*, 1959-1967; *Homer Tribune*, 1991 to present.

"A crew of 21 men are working out of Ninilchik doing oil exploration within the Caribou Hills," *Homer News*, Dec. 5, 1968.

"Agencies admit to failing to protect water sources from fuel pollution," by Judy Cart, *Los Angeles Times*, March 10, 2015, http://www.latimes.com/local/california/la-me-fracking-water-20150311-story.html.

"Alaska LNG Project buying land, securing access," by Tim Bradner, *Alaska Journal of Commerce*, July 31, 2014, http://www.alaskajournal.com/business-and-finance/2014-07-31/alaska-lng-project-buying-land-securing-access#.Vtgxh5MrJPM.

"Among Hammond's many accomplishments was protecting Kachemak Bay's resources," Point of View by Loren Flagg, *Homer News*, Aug. 11, 2005.

"Analyst warns of gas shortage for Cook Inlet," by Alan Bailey, *Alaska Dispatch News*, Oct. 20, 2012, http://www.adn.com/article/20121020/analyst-warns-gas-shortage-cook-inlet.

"Anchor Point offshore drilling closely watched," *Homer News*, Aug. 17, 1967.

"ANWR wilderness is vital to caribou, Gwich'in people in Alaska," Commentary by Trimble Gilbert, *Alaska Dispatch News*, March 13, 2015, http://www.adn.com/article/20150313/anwr-wilderness-vital-caribou-gwichin-people-alaska.

"Armstrong, ENSTAR agree to send North Fork gas to Homer,"
by Aaron Selbig, *Homer News*, Nov. 14, 2008.

"As Oil Money Melts, Alaska Mulls First Income Tax in 35 Years,"
by Kirk Johnson, *The New York Times*, Dec. 25, 2015,
http://www.nytimes.com/2015/12/26/us/as-oil-money-melts-alaska-
mulls-first-income-tax-in-35-years.html.

"Atlantic Richfield producing at Prudhoe," *Homer News*, Oct. 23, 1969.

"Atlantic Seismic Tests For Oil: Marine Animals at Risk?"
By Helen Scales, *National Geographic*, March 1, 2014, http://news.
nationalgeographic.com/news/energy/2014/02/140228-atlantic-
seismic-whales-mammals/.

"Bill stirs reaction," *Homer News*, March 8, 1977.

"BlueCrest, WesPac ink deal to develop Cosmo," by Tim Bradner, *Alaska
Journal of Commerce*, Jan. 15, 2015, http://www.alaskajournal.com/
business-and-finance/2015-01-15/bluecrest-wespac-ink-deal-develop-
cosmo#.Vtgc1ZMrKb8.

"Borough officials hold Nikiski LNG meeting," by Ben Boettger, *Peninsula
Clarion*, Nov. 14, 2015, http://peninsulaclarion.com/news/2015-11-14/
borough-officials-hold-nikiski-lng-meeting.

"Buccaneer bankruptcy demand hits Homer businesses, agencies,"
by Michael Armstrong, *Homer News*, July 16, 2015.

"The Caribou Question, the caribou and Alaska oil," by Deborah Jacobs,
Property and Environment Research Center, PERC Report: Volume 19,
No. 2, Summer 2001, http://www.perc.org/articles/caribou-question

"CINGSA to sell its newly discovered gas," by Ben Boettger, *Peninsula
Clarion*, Dec. 6, 2015, http://peninsulaclarion.com/news/2015-12-06/
cingsa-can-sell-some-of-its-gas.

"Company says Cook Inlet oil well tests exceed expectations," Associated
Press, *Fairbanks Daily News-Miner*, Dec. 31, 2013, http://www.
newsminer.com/business/company-says-cook-inlet-well-tests-exceed-
expectations/article_fbbd94e2-7245-11e3-86c1-001a4bcf6878.html.

"Company seeks permits to extend well to pipeline,"
by Marcus K. Garner, *Peninsula Clarion*, Oct. 5, 2003.

"Construction gets under way on Cosmopolitan development,"
by Michael Armstrong, *Homer News*, June 17, 2015

"Council passes gas roll," by Michael Armstrong, *Homer News*,
March 25, 2015.

"Court says CIRI can push for Kenai gas royalty payments," by Elwood
Brehmer, *Alaska Journal of Commerce*, Aug. 21, 2014, http://www.

alaskajournal.com/business-and-finance/2014-08-21/court-says-ciri-can-push-kenai-gas-royalty-payments#.VtdG-ZMrJPM.

"Dead ducks," *Homer News*, Nov. 30, 1967.

"Deep Creek well at 13,500 ft, Richfield terms second Swanson well productive," *Homer Herald*, June 12, 1958.

"Discussion begins on oil tax credits," by Tim Bradner, Morris News Service-Alaska, published in the *Homer News*, July 16, 2015.

"Drill rig in bay," *Homer News*, April 1, 1964.

"Drilling platform for Cook Inlet oil field," *Homer News*, June 10, 1965.

"Eminent Domain Meeting in Nikiski," KSRM 920 Radio Online, http://www.radiokenai.us/eminent-domain-meeting-in-nikiski/.

"Exploration well to be drilled on North Fork Road," *Homer News*, Dec. 29, 1966.

"Federal judge rejects Parnell-era push to open ANWR," by Lisa Demer, *Alaska Dispatch News*, July 21, 2015, http://www.adn.com/article/20150721/federal-judge-rejects-parnell-era-push-open-anwr.

"Feds cancel offshore lease sales in Arctic," by Erica Martinson, *Alaska Dispatch News*, Oct. 17, 2015.

"Feds OK drilling permit in Arctic reserve," by Alex DeMarban, *Alaska Dispatch News*, Oct. 23, 2015.

"Ferris arrives in Homer," *Homer Weekly News*, Feb. 6, 1975.

"Fishermen protest inlet waste," *Homer News*, Nov. 9, 1967.

"For the sake of the Inupiat, Shell should give up drilling in the Arctic," Commentary by Othniel Art Oomittuk Jr., *Alaska Dispatch News*, Aug. 10, 2015, http://www.adn.com/node/2765381.

"Gov. Bill Walker: Alaska's budget challenge, tightening our belts," *Alaska Dispatch News*, Dec. 8, 2015, http://www.adn.com/print/article/20151208/gov-bill-walker-alaskas-budget-challenge-tightening-our-belts.

"Gov. Walker: Enough is enough on ANWR," Community Perspective by Gov. Bill Walker, *Fairbanks Daily News-Miner*, Feb. 1, 2015, http://www.newsminer.com/opinion/community_perspectives/gov-walker-enough-is-enough-on-anwr/article_4f751afe-aab4-11e4-b0d1-17f146149054.html.

"Gov's plan aims to reshape state's relationship with oil," by Rachel Waldholz, Alaska Public Radio Network, Jan. 18, 2016, http://www.ktoo.org/2016/01/18/123434/.

"Head of California agency accused of favoring oil industry quits," by Julie Cart, *Los Angeles Times*, June 5, 2015, http://www.latimes.com/local/lanow/la-me-head-of-oil-regulating-agency-quits-20150605-story.html.

"Hilcorp acquires more inlet oil assets," by Tim Bradner, Morris News Service-Alaska, published in *Homer News*, July 9, 2015.

"Hilcorp applies to drill 2 wells," by Elizabeth Earl, Morris News Service-Alaska, published in *Homer News*, Sept. 17, 2015.

"Hilcorp goes big, major gas exploration program follows disappointing oil exploration results," by Eric Lidji for *Petroleum News*, week of April 30, 2014, http://www.petroleumnews.com/pntruncate/22643877.shtml.

"Habitat approved for drilling; fishermen-environmentalists on collision course with oil interests," *Homer Weekly News*, Nov. 14, 1974.

"Homer lawmaker introduces bill creating Alaska income tax as aid in countering deficit," by Nathaniel Herz, *Alaska Dispatch News*, April 3, 2015, http://www.adn.com/article/20150403/homer-lawmaker-introduces-bill-creating-alaska-income-tax-aid-countering-deficit.

"Homer takes oil stand," *Homer Weekly News*, Oct. 10, 1974.

"Inlet View Lodge celebrates 20 years with same owner," by McKibben Jackinsky, *Homer News*, April 28, 2010.

"'It's just too big a prize': Why Shell sticks to Chukchi plans, despite obstacles," *Alaska Dispatch News*, Aug. 5, 2015, https://www.adn.com/article/20150805/its-just-too-big-prize-why-shell-sticks-chukchi-plans-despite-obstacles.

"Kachemak Bay defense fund started," *Homer Weekly News*, Nov. 21, 1974.

"Kenai Peninsula Fair: 63 years old and counting," by McKibben Jackinsky, *Homer News*, Aug. 13, 2014.

"Kenai pipeline up and running," by Tim Bradner, Morris News Service, *Homer News*, Sept. 11, 2003.

"Kenai pipeline will connect new gas source," Associated Press, published in *Peninsula Clarion*, Jan. 29, 2002.

"LNG Project Meetings in Anchorage, Fairbanks," KSRM 920 Radio Online, July 11, 2014, http://radiokenai.net/lng-project-meetings-in-anchorage-fairbanks/.

"Locate coal and oil at Kachemak Bay," *Seward Weekly Gateway*, Feb. 9, 1907, from A Collection of Historical Newspaper Articles, 1895-1913, compiled by David Brann, Homer, Alaska.

"Largest oil well in Alaska!" *Homer News*, July 1, 1965.

"Let the People Speak," Visitor Voices by Mike O'Meara, *Journal of Museum Education*, vol. 28, number 3, Fall 2003, http://www.jstor. org/stable/40479303?seq=1#page_scan_tab_contents.

"Marathon to expand Susan Dionne pad," by Kristen Nelson, *Petroleum News*, Vol. 13, No. 20, week of May 18, 2008, http://www. petroleumnews.com/pntruncate/291305330.shtml.

"Micciche returns to Soldotna for town hall meeting," by Dan Balmer, *Peninsula Clarion*, March 9, 2014, http://peninsulaclarion.com/ news/2014-03-09/micciche-returns-to-soldotna-for-town-hall-meeting.

"Murkowski fears Obama may unilaterally declare ANWR a national monument," by Rod Boyce, *Fairbanks Daily News-Miner*, Jan. 25, 2015, http://www.newsminer.com/news/local_news/murkowski-fears-obama-may-unilaterally-declare-anwr-a-national-monument/article_ c58971aa-a4dc-11e4-9436-f33f6df1c48a.html.

"New oil well at Ninilchik announced," *Homer Herald*, Feb. 20, 1958.

"Ninilchik beach closed to clam digging," by McKibben Jackinsky, *Homer News*, March 19, 2014.

"Notice of Competitive Oil and Gas Lease Sale No. 19," published in *Homer News*, March 2, 1967.

"Obama to seek wilderness designation for Alaska refuge," by Rod Boyce, *Fairbanks Daily News-Miner*, Jan. 25, 2015, http://www.newsminer. com/news/alaska_news/obama-to-seek-wilderness-designation-for-alaska-refuge/article_c75b6918-a4b7-11e4-82d1-1be011fdb032.html.

"Oil: Wildcatting v. Wildlife," *Time*, Dec. 16, 1957, http://www.time.com/ time/magazine/article/0,9171,893828,00.html.

"Oil production in Alaska is declining," a message from Alaska oil and gas industry, *Homer News*, March 12, 1992.

"Oil companies say they hope to avoid using eminent domain on Nikiski land," by Dermot Cole, *Alaska Dispatch News*, Oct. 4, 2014, http:// www.adn.com/article/20141004/ oil-companies-say-they-hope-avoid-using-eminent-domain-nikiski-land.

"Oil in Alaska to stay," *Homer Weekly News*, Feb. 19, 1976.

"Oil Prices: What's Behind the Drop? Simple Economics," by Clifford Krauss, *The New York Times*, updated Jan. 22, 2016, http://www.nytimes.com/interactive/2016/business/energy-environment/oil-prices.html?_r=1.

"Oil reported in Swanson River Well #3," *Homer Herald*, Oct. 2, 1958.

"Oil rig arrives here," *Homer News*, Nov. 17, 1966.

"Oil seepage is regarded most important find," *Seward Gateway*, July 18, 1936, from A Collection of Historical Newspaper Articles, 1895-1913, compiled by David Brann, Homer, Alaska.

"Petroleum industry service boats are now coming to Homer for fuel, because they can also obtain water," *Homer News*, July 25, 1968.

"Producers 2013: Hilcorp: biggest little newcomer," by Eric Lidji, *Petroleum News*, Vol. 18, No. 46, week of Nov. 17, 2013, http://www.petroleumnews.com/pntruncate/788709873.shtml.

"Resource committee passes buyback bill," *Homer News*, April 7, 1977.

"Remote Alaska fishing town braces for Arctic oil development," by Lauren Rosenthal, Aljazeera America, Jan. 15, 2014, http://america.aljazeera.com/articles/2014/1/15/remote-alaskan-fishingtownbracesforarcticoildevelopment.html.

"Scientists are floored by what's happening in the Arctic right now," by Chris Mooney, *The Washington Post*, Feb. 18, 2016, https://www.washingtonpost.com/news/energy-environment/wp/2016/02/18/scientists-are-floored-by-whats-happening-in-the-arctic-right-now/.

"Scouting review," *Homer News*, Nov. 16, 1967.

"Shadow Over An Ancient Land," by Susan Reed, *People*, Sept. 18, 1989, Vol. 32, No. 12, http://www.people.com/people/archive/article/0,,20121200,00.html.

"Shell move dims oil prospects, delights environmentalists," by Dan Joling and Jonathan Fahey, Associated Press, Sept. 27, 2015, http://www.seattletimes.com/nation-world/shell-says-it-will-cease-alaska-offshore-arctic-drilling/.

"Standard Oil Company to drill in Naptowne," *Homer News*, Jan. 6, 1966.

"Standard Oil Company Will Develop Property at Cook Inlet," June 10, 1899, *Seattle Post-Intelligencer*, from A Collection of Historical Newspaper Articles, 1895-1913, compiled by David Brann, Homer, Alaska.

"Standard Oil to drill Beluga Well," *Homer News*, March 24, 1966

"State needs to be cautious when extending its 'open for business' invitation to anyone," Point of View by Bob Shavelson, executive director of Cook Inletkeeper, printed in *Homer News*, Feb. 20, 2014, http://homernews. com/homer-opinion/point-of-view/2014-02-20/ state-needs-to-be-cautious-when-extending-its-%E2%80%98open-for.

"State team assesses oil reservoir potential on west side of Cook Inlet," by Alan Bailey, *Petroleum News*, printed in *Alaska Dispatch News*, Aug. 1, 2015, https://www.adn.com/article/20150801/ state-team-assesses-oil-reservoir-potential-west-side-cook-inlet.

"State, local leaders discuss PILT split from AK LNG Project," by Elwood Brehmer, *Alaska Journal of Commerce*, Dec. 22, 2015, http://www. alaskajournal.com/2015-12-22/state-local-leaders-discuss-pilt-split-ak-lng-project#.VtgxCZMrKb8.

"Study Raises Concerns About Toxic Oil and Gas Emissions in California," by David Hasemyer, *InsideClimate News*, Jan. 26, 2015, http:// insideclimatenews.org/news/20150126/ study-raises-concerns-about-toxic-oil-and-gas-emissions-california-air.

"Texaco Inc. makes application for discovery well," *Homer News*, June 30, 1966.

"Thanks for work to advance gas line," Commentary by Gov. Bill Walker, *Alaska Dispatch News*, Nov. 8, 2015.

"This Company Gave Every One of Its Employees a $100k Christmas Bonus," by Chris Matthews, *Fortune*, Dec. 11, 2015, http://fortune. com/2015/12/11/100k-christmas-bonus/.

"Tim Bradner: Southcentral faces running short of natural gas," by Tim Bradner, *Alaska Journal of Commerce*, published in *Alaska Dispatch News*, Oct. 20, 2012, http://www.adn.com/article/20121020/ tim-bradner-southcentral-faces-running-short-natural-gas.

"Too early to answer: Nikiski residents have questions about proposed LNG plant," Kaylee Osowski, *Peninsula Clarion*, May 14, 2014, http:// peninsulaclarion.com/news/2014-05-13/too-early-to-answer.

"U.N. Weather Agency: It's Record Hot Out There This Year," Associated Press, NBC News, Nov. 25, 2015, http://www.nbcnews.com/tech/ tech-news/u-n-weather-agency-its-record-hot-out-there-year-n469456.

"Union, Ohio Oil to drill soon, peninsula site not announced by company," *Homer Herald*, July 10, 1958.

"Unocal hopes to find gas reserves near Ninilchik, Clam Gulch," by Doug Loshbaugh, *Alaska Journal of Commerce*, May 5, 2001, http://www.

alaskajournal.com/community/2001-05-06/unocal-hopes-find-gas-reserves-near-ninilchik-clam-gulch#.VthYiZMrKb9.

"'A very sad day for Alaska': State leaders react as Shell ceases exploration," by Rebecca Palsha, KTUU Channel 2 Anchorage, Sept. 28, 2015, http://www.ktuu.com/news/news/a-very-sad-day-for-alaska-state-leaders-react-as-shell-ceases-exploration/35532108.

"Welcome to the neighborhood: North Fork residents adjust to natural gas development," by McKibben Jackinsky, *Homer News*, April 20, 2011.

"Westward Ho! Coal and Oil as Well as Gold," July 20, 1901, *The Alaskan*, Sitka, from A Collection of Historical Newspaper Articles, 1895-1913, compiled by David Brann, Homer, Alaska.

"What will gas cost?" by McKibben Jackinsky, *Homer News*, Nov. 1, 2012.

"Wildcat progressing," *Homer News*, Aug, 5, 1965.

PRESS RELEASES:

"ConocoPhillips Q3 Alaska Earnings Report Shows Profits of $5,369,565 per day, or $22,731 per hour," press release from Sen. Bill Wielechowski, Nov. 1, 2013, http://alaskasenatedems.com/wielechowski/2013/11/01/news-conocophillips-q3-alaska-earnings-report-shows-profits-of-5369565-per-day-or-22731-per-hour/.

"Cook Inlet oil royalties on the rise," press release from the Alaska Department of Natural Resources, Aug. 14, 2014.

"Governor releases amended endorsed budget," press release from Governor's Office, Feb. 5, 2015.

"Senate Majority Encouraged by Alaska LNG Roadmap Agreements Reached by Parnell Administration," press release from Alaska Senate Majority, Jan. 15, 2014.

"State Reaches Settlement on Seabulk Pride Spill and Grounding," press release from Alaska Department of Law, July 1, 2010, http://law.alaska.gov/press/releases/2010/070110-SeabulkPride.html..

"State signs commercial agreements for Alaska LNG Project," press release from Alaska Department of Natural Resources, Jan. 15, 2014.

Public Records

The Alaska Department of Natural Resources, Recorder's Office, is the source for vital statistics data as well as information on individual oil and gas leases and other public records.

Of personal interest to me were two files on my grandfather's Ninilchik homestead, "United States of America Patent for 160-acre homestead, Walter Jackinsky Sr., Jan. 12, 1937," and "United States of America Patent, Walter Jackinsky Sr., April 14, 1952." Over the years several Jackinsky family members, my father, Walter Jackinsky Jr., among them, have individually signed agreements to allow oil and gas development on their inherited portions of the original homestead. According to records available on the Recorder's Office website, three have signed leases with Hilcorp Alaska. http://dnr.alaska.gov/ssd/recoff/default.cfm,

Letters:

Letter from the President – Arctic National Wildlife Refuge Proposed Designations, April 3 2015, https://www.whitehouse.gov/the-press-office/2015/04/03/letter-president-arctic-national-wildlife-refuge-proposed-designations.

Letter to Larry Burgess, BlueCrest Energy Inc. regarding LOCI 14-007 Hansen Pad, Cosmopolitan Project, Lease Plan of Operations Decision, from the Alaska Department of Natural Resources, Division of Oil and Gas, Feb. 6, 2015.

Letter to Kathy Foerster, Chair, Alaska Oil and Gas Conservation Commission regarding application for spacing exception per 20 AAC 25.055(a)(2) and (4) to test and produce the Paxton #6 within the Ninilchik Unit Undefined Tyonek and Beluga Pools, from Hilcorp Alaska, LLC, June 24, 2014.

Letter to Jolie Harrison of the National Marine Fisheries Service regarding authorization for incidental take of marine mammals during seismic survey in Alaska's Cook Inlet, March 1, 2015, to Feb. 29, 2020, from the Resource Development Council for Alaska Inc., April 3, 2015, http://www.akrdc.org/cook-inlet-incidental-take-authorizations-2015---2020.

Letter to Brian Havelock, Division of Oil and Gas, Alaska Department of Natural Resources regarding Armstrong Cook Inlet, LLC, LO/CI 10-004, from the Kenai Peninsula Borough, Aug. 30, 2010.

Letter to Landowner, received by McKibben Jackinsky, regarding Miscellaneous Land Use Application to conduct geophysical activities on submerged lands adjacent to landowner's property, from Sue Simonds, SAExploration, July 7, 2015.

Letter to McKibben Jackinsky, Jennifer and Emily Long regarding public awareness safety, from ENSTAR Natural Gas Company, March 28, 2007.

Letter to McKibben Jackinsky, Jennifer and Emily Long regarding Cosmopolitan Pipeline courtesy notice, from NORSTAR Pipeline Company, June 5, 2007.

Letter to McKibben Jackinsky, Emily Long and Jennifer Long regarding oil and gas lease proposal, from Hilcorp Alaska LLC, June 24, 2013.

Letter to McKibben Jackinsky, Jennifer and Emily Long regarding oil and gas lease proposal, from Hilcorp Alaska LLC, Aug. 29, 2013.

Letter to Mildred M. Martin regarding EDCON-Unocal Alaska 2003 Kenai Peninsula Regional Gravity Survey from EDCON, Jan. 5, 2004.

Letter to Ed Teng, Armstrong Cook Inlet LLC, regarding Unit Plan of Operations Approval, North Fork Unit Gas Development, Sept. 29, 2010.

Letters of Objections Filed Through Tuesday January 22, 2013, Ordinance 13-02, Creating the Natural Gas Distribution Special Assessment District, City of Homer.

WEBSITES

A Layman's Guide for Establishing a Utility Special Assessments District (USAD), Kenai Peninsula Borough Assessing Department, http://www.kpb.us/images/KPB/ASG/Documents/LAYMANS_GUIDE_USAD.pdf.

A Sense of the Refuge, Arctic National Wildlife Refuge, U.S. Fish and Wildlife Service, October 2011, http://www.fws.gov/uploadedFiles/Region_7/NWRS/Zone_1/Arctic/PDF/Sense%20of%20the%20Refuge%20booklet%20web.pdf.

Alaska Department of Natural Resources, Division of Oil and Gas, 2014 Annual Report, http://dog.dnr.alaska.gov/publications/documents/annualreports/2014_annual_report.pdf.

Alaska Housing Finance Corporation, https://www.ahfc.us/.

Alaska Volcano Observatory, http://www.avo.alaska.edu/.

"ANWR Reality Lies Far North of Gwich'in," by George Tagarook, http://anwr.org/category/people/.

"AOGCC, 50 Years of Service to Alaska," published by the Alaska Oil and Gas Conservation Commission, revised Oct. 10, 2010, http://doa.alaska.gov/ogc/WhoWeAre/50th/aogcc50thBooklet.pdf .

AOGCC Pool Statistics, Swanson River Field, Hemlock Oil Pool, Alaska Department of Administration, Alaska Oil and Gas Conservation Commission, http://doa.alaska.gov/ogc/annual/2004/Oil_Pools/ Swanson%20River%20-%20Oil/1_Oil_1.html.

Apache Corporation, www.apachecorp.com.

"Arctic National Wildlife Refuge, an irreplaceable natural treasure," *Audubon*,

https://www.audubon.org/conservation/arctic-national-wildlife-refuge.

Arctic National Wildlife Refuge Comprehensive Conservation Plan, "Dear Reader," http://www.fws.gov/home/arctic-ccp/pdfs/Final_DearReader_ Ltr_2015.pdf.

Arctic National Wildlife Refuge, Overview of Refuge Purposes, U.S. Fish and Wildlife Service, http://www.fws.gov/refuge/arctic/purposes.html.

Arctic National Wildlife Refuge, U.S. Fish and Wildlife Service, Planning Update 5, 2015, https://www.fws.gov/home/arctic-ccp/pdfs/Update%20 5%20merged%202015%20(1).pdf.

Arctic Refuge and the Wilderness Act: There's a Connection, U.S. Fish and Wildlife Service, http://www.fws.gov/refuge/arctic/ wildernessconnection.html.

BlueCrest Energy, bluecrestenergy.com.

Bureau of Mines, Minerals yearbook area reports: domestic 1975, page 59, http://digicoll.library.wisc.edu/cgi-bin/EcoNatRes/EcoNatRes- idx?type=turn&entity=EcoNatRes.MinYB1975v2. p0067&id=EcoNatRes.MinYB1975v2&isize=M.

Caribou (Rangifer tarandus granti), Research, Alaska Department of Fish and Game, www.adfg.alaska.gov/index.cfm?adfg=caribou.research.

Caribou Management Report, Porcupine Herd, From July 1, 2012 to June 30, 2014, Alaska Department of Fish and Game, Division of Wildlife Conservation, http://www.adfg.alaska.gov/static/research/wildlife/ speciesmanagementreports/pdfs/caribou_2015_chapter_15_porcupine. pdf.

Central and Porcupine Herds Survey Results, Nov. 23, 2014, http://anwr. org/2014/11/central-and-porcupine-herds-survey-results/.

Cook Inlet Oil and Gas Activity, State of Alaska, Department of Natural Resources, Division of Oil and Gas, as of November 2015, http://dog. dnr.alaska.gov/GIS/Data/ActivityMaps/CookInlet/ CookInletOilAndGasActivityMap-201511.pdf. (Basic link: http://dog.dnr.alaska.gov/GIS/ActivityMaps.htm.)

Cook Inletkeeper, http://inletkeeper.org/.

Deerstone Consulting, http://deerstoneconsulting.com/.

"'Drill, baby, drill!' almost didn't happen," by Josh Kurtz, Climate Wire, E&E Publishing LLC, Aug. 29, 2012, http://www.eenews.net/stories/1059969331.

"Drilling for oil on Alaska's Kenai Peninsula," Alexander's Gas and Oil Connection, an Institute for Global Energy Research, May 4, 2003, http://www.gasandoil.com/news/n_america/6e10d365d7f83ffe5b74d0ca0bc54925.

Fire Island Wind, http://fireislandwind.com/.

"Fracking Threatens Health of Kern County Communities Already Overburdened with Pollution," Fact Sheet, Natural Resource Defense Council, September 2014, http://www.nrdc.org/health/files/california-fracking-risks-CA-FS.pdf.

"Frequently asked questions – How much oil is consumed in the United States," U.S. Energy Information Administration, http://www.eia.gov/tools/faqs/faq.cfm?id=33&t=6.

Homer Electric Association, www.homerelectric.com.

Kachemak Bay Conservation Society, www.kbayconservation.org.

Kachemak Heritage Land Trust, www.kachemaklandtrust.org.

Kate Boyan's Gallery, http://livingbeadwork.blogspot.com/.

Kenai Peninsula Borough Ordinance 2000-02 (Popp, Sprague) Substitute, An Ordinance Repealing KPB 21.08 "Local Option" and Adopting KPB 21.44 "Local Option Zoning," May 16, 2000, http://www.kpb.us/assembly-clerk/legislation/ordinances

Kenai Peninsula Borough, Alaska, Code of Ordinances, Chapter 21.44 – Local Option Zoning, https://www.municode.com/library/ak/kenai_peninsula_borough/codes/code_of_ordinances?nodeId=TIT21ZO_CH21.44LOOPZO.

Kenai National Wildlife Refuge, About the Refuge, U.S. Fish and Wildlife Service, http://www.fws.gov/refuge/Kenai/about.html.

Kenai National Wildlife Refuge, Oil and Gas Assessment, By Ronald Teseneer, Christopher Gibson, Aden Seidlitz and James Borkowski, U.S. Department of the Interior Bureau of Land Management, http://www.blm.gov/style/medialib/blm/ak/aktest/energy/energy_publications.Par.78078.File.dat/kenai_oga.pdf.

Kenai Peninsula State Fair 2014 Annual Report, www.kenaipeninsulafair. com (under "2014 Snapshot").

"Lakeview Gusher: A brief history," by Jonathan Montgomery, May 1, 2011, http://www.counterspill.org/article/lakeview-gusher-brief-history.

Land Trust Alliance, www.landtrustalliance.org.

"Let the People Speak," guest blog for Cook Inletkeeper by Mike O'Meara, March 24, 2010, https://inletkeeper.org/blog/let-the-people-speak.2.

Memorandum Decision and Order, State of Alaska, plaintiff, v. Sally Jewell, in her official capacity as the United States Secretary of the Interior, et al., defendants, and Gwich'in Steering Committee, et al., intervenor-defendants, Case No. 3:14-cv-00048-SLG, United States District Court for the District of Alaska, http://www.trustees.org/wp-content/ uploads/2015/07/2015-07-21-65-Decision-and-Order.pdf.

Moore v. State, No. 3551 (553 P. 2d 8) (Alaska July 9, 1976), Environmental Law Reporter, https://elr.info/sites/default/files/ litigation/6.20813.htm.

National Renewable Energy Laboratory, http://www.nrel.gov/.

National Wildlife Refuges, Improvement Needed in the Management and Oversight of Oil and Gas Activities on Federal Lands, Testimony Before the Subcommittee on Fisheries Conservation, Wildlife, and Oceans, Committee on Resources, Statement of Barry T. Hill, Director, Natural Resources and Environment, House of Representatives, United States General Accounting Office, Oct. 30, 2003, http://www.gao.gov/ assets/120/110484.pdf.

Ninilchik Unit map, Division of Oil and Gas, Alaska Department of Natural Resources, http://dog.dnr.alaska.gov/Units/Documents/ UnitMaps/CookInlet/NinilchikUnitMap-201509.pdf.

"Obama Is Trying To Protect A Huge Arctic Wildlife Zone, But Congress Likely Won't Have It," by Natasha Geiling, *Think Progress*, April 6, 2015, http://thinkprogress.org/climate/2015/04/06/3643159/ anwr-protections-finalized-obama/.

Oil and Gas: Incidental Take Authorizations, NOAA Fisheries, National Oceanic and Atmospheric Administration, http://www.nmfs.noaa.gov/ pr/permits/incidental/oilgas.htm

Petroleum Systems and Geologic Assessment of Oil and Gas in the San Joaquin Basin Province, California, edited by Allegra Hosford Scheirer, U.S. Geological Survey Professional Paper 1713, 2007, http://pubs.usgs. gov/pp/pp1713/.

Potential seismic hazards and tectonics of the upper Cook Inlet basin, Alaska, based on analysis of Pliocene and younger deformation, Geological Society of America Bulletin, by P.J. Haeussler, R.L. Bruhn and T.L. Pratt, https://pubs.er.usgs.gov/publication/70022145.

Pratt Museum, Darkened Waters: Profile of an Oil Spill, http://www. exhibitfiles.org/dfile2/ReviewWalkthrough/219/original/DARKENED. PDF.

"President Obama Calls on Congress to Protect Arctic Refuge as Wilderness," by John Podesta and Mike Boots, The White House Blog, Jan. 25, 2015, https://www.whitehouse.gov/blog/2015/01/25/ president-obama-calls-congress-protect-arctic-refuge-wilderness.

Protecting the Arctic National Wildlife Refuge, National Wildlife Refuge Association, http://refugeassociation.org/advocacy/refuge-issues/arctic/.

Public Land Order 2214, Establishing the Arctic National Wildlife Refuge, U.S. Fish and Wildlife Service, http://www.fws.gov/refuge/ arctic/plo2214.html.

Resolution 03-017(S) of the City Council of Homer, Alaska, Approving A Contribution in Aid of Construction Agreement (CIAC) Between the City of Homer and ENSTAR Natural Gas Company for Construction of Improvements Within the Homer Natural Gas Distribution System Special Assessment District, In An Amount Not To Exceed $12,160,632, http://www.cityofhomer-ak.gov/resolution/ resolution-13-017s-approving-contribution-aid-construction- agreement-ciac-between-city.

Revised Comprehensive Conservation Plan and Final Environmental Impact Statement for the Arctic National Wildlife Refuge, U.S. Department of the Interior, Fish and Wildlife Service, Arctic National Wildlife Refuge, Volume 1, January 2015, http://www.fws.gov/home/ arctic-ccp/pdfs/00_FrontPages_Vol1_Jan2015_web.pdf.

Revised Comprehensive Conservation Plan and Final Environmental Impact Statement for the Arctic National Wildlife Refuge, U.S. Department of the Interior, Fish and Wildlife Service, Arctic National Wildlife Refuge, Volume 2 – Appendices, January 2015, http://www. fws.gov/home/arctic-ccp/pdfs/01_FrontPages_Vol2_Jan2015_web.pdf.

Revised Comprehensive Conservation Plan and Final Environmental Impact Statement for the Arctic National Wildlife Refuge, U.S. Department of the Interior, Fish and Wildlife Service, Arctic National Wildlife Refuge, Volume 3 – Response to Public Comments, January 2015, http://www.fws.gov/home/arctic-ccp/pdfs/01_Vol3_FrontPages_ Jan2015_web.pdf.

Revised Comprehensive Conservation Plan and Final Environmental Impact Statement for the Arctic National Wildlife Refuge, U.S. Department of the Interior, Fish and Wildlife Service, Arctic National Wildlife Refuge, Volume 4 – Sample of Public Comments, January 2015, http://www.fws.gov/home/arctic-ccp/pdfs/00_Vol4_FrontPages_Jan2015_web.pdf.

The 10 Biggest Oil Spills in World History – 2 Lakeview Gusher Number One, Oil and Gas IQ, The 10 Biggest Oil Spills In World History - 2 Lakeview Gusher Number One | Oil and Gas IQ | Upstream & Downstream Oil and Gas Industry News & Information.

Urban Green Technology, http://urbangreentechnology.com/.

"What are conservation easements?" The Nature Conservancy, http://www.nature.org/about-us/private-lands-conservation/conservation-easements/what-are-conservation-easements.xml.

The Wilderness Act, Public Law 88-577 (16 U.S.C 1131-1136), 88th Congress, Second Session, September 3, 1964, http://www.wilderness.net/nwps/legisact.

"Wilderness Stewardship: Protecting Wilderness Character," U.S. Fish and Wildlife Service, Arctic National Wildlife Refuge, http://www.fws.gov/refuge/arctic/wildernessstewardship.html.

Wisdom and Associates, Inc., http://www.wisdomandassociates.com/index.html.

OTHER:

A Moral Choice for the United States: The Human Rights Implications for the Gwich'in of Drilling in the Arctic National Wildlife Refuge, by the Gwich'in Steering Committee, The Episcopal Church, Richard J. Wilson, professor of Law and Director of the International Human Rights Law Clinic at American University, 2005.

A Seismic Attribute Study to Assess Well Productivity In the Ninilchik Field, Cook Inlet Basin, Alaska, A Thesis Submitted to the Graduate Faculty of the Louisiana State University and Agricultural and Mechanical College, by Andrew Sampson, May 2012.

Alaska Economic Update – Part 4, by Mark Edwards, alaskanomics.com, Northrim Bank, http://www.alaskanomics.com/2015/03/alaska-economic-update-part-4.html.

Alaska Oil Production, Production History FY* 1959-2012, http://www.tax.alaska.gov/sourcesbook/AlaskaProduction.pdf.

Alaska Statutes Section 09.55.240: Uses for which authorized; rights-of-way.

An Introduction to Alaska Fiscal Facts and Choices, by Gunnar Knapp, director and professor of economics, Institute of Social and Economic Research, University of Alaska Anchorage, June 5, 2015, http://www.iser.uaa.alaska.edu/Publications/presentations/2016_02_02-AnIntroductionAKFiscaFactsChoices.pdf.

"ANWR decision is wrong, offensive," Alaska Oil and Gas Association, Jan. 25, 2015.

"Apache Honored For Stewardship in Alaska," Nov. 5, 2013, http://www.adn.com/article/20131112/apache-corp-honored-stewardship-alaska.

Application of Hilcorp Alaska, LLC to modify the well spacing and escrow account requirements of Conservation Order No. 701, Beluga/Tyonek Gas Pool, Ninilchik Unit, onshore and offshore Kenai Peninsula Borough, State of Alaska, Alaska Oil and Gas Conservation Commission, Docket Number: CO-14-029, Conservation Order No. 701A, Ninilchik Unit, Ninilchik Field, Beluga/Tyonek Gas Pool, Kenai Peninsula Borough, Alaska, Oct. 9, 2015.

"Artist Stops Oil Pipeline Cold," by occupystephanie, Daily Kos, May 30, 2014, http://www.dailykos.com/story/2014/05/30/1303087/-Artist-Stops-Oil-Pipeline-Cold.

Cook Inlet Geographic Response Strategies, April 2003.

"Cook Inlet pipeline crossing is about making the best choices," by Larry Persily, Kenai Peninsula Borough Mayor's Office, Aug. 26, 2015.

Deed of Conservation Easement, Michael S. and Janet V. O'Meara, and John S. Bogel, and Kachemak Heritage Land Trust, Dec. 27, 1991.

Directional and Horizontal Drilling in Oil and Gas Wells, http://geology.com/articles/horizontal-drilling/.

District 31 Constituent Surveys, State of Alaska – Revenue Options, Office of Representative Paul Seaton, Oct. 12, 2015.

Evaluation of the 2009 Drift River Oil Terminal Coordination and Response with a Review of the Cook Inlet RCAC's Role in Oil Spill Response, prepared by Pearson Consulting LLC, commissioned by the Cook Inlet RCAC, June 2010, http://www.circac.org/wp-content/uploads/finaldriftriver.pdf.

Fact Sheet: Agricultural Land for Alaska, Alaska Department of Natural Resources, Division of Agriculture.

Gas Contingency Planning, Railbelt electric and gas utilities harmonize efforts with local governments, cooperative effort by Municipality of Anchorage, Kenai Peninsula Borough, Matanuska-Susitna Borough, ENSTAR Natural Gas Company, Chugach, Homer Electric Association Inc., Municipal Light and Power, Matanuska Electric Association; Oct. 11, 2013.

Geology of the Iniskin-Tuxedni Region, Alaska, by Robert L. Detterman and John K. Hartsock, Geological Survey Professional Paper 512, U.S. Department of the Interior, 1966.

"Inletkeeper (Still) Fishing to End Toxic Oil and Gas Loophole," *Inletkeeper*, distributed by Cook Inletkeeper, Winter 2013-2014.

Kenai-Kachemak Pipeline Project Design Basis and Criteria, prepared by Kenai-Kachemak Pipeline LLC, July 2002.

"Know Your Property Rights!", Cook Inletkeeper, http://inletkeeper.org/ resources/contents/oilgaspropertyrights/view.

Land Access Agreement through Dec. 31, 2014, request to McKibben Jackinsky, Jennifer and Emily Long from SAExploration.

"LNG plant construction a huge undertaking," by Larry Persily, Kenai Peninsula Borough Mayor's Office, Sept, 16, 2015.

"More than 1/3 of Alaska's jobs are tied to the oil and gas industry," Facts and Figures, Alaska Oil and Gas Association, http://www.aoga. org/facts-and-figures.

Negative Secondary Impacts from Oil and Gas Development, The Energy & Biodiversity Initiative, http://www.theebi.org/pdfs/impacts.pdf.

"Not just Exxon: The Entire Oil and Gas Industry Knew About Climate Change 35 Years Ago," Daily Kos, Dec. 25, 2015, https://m.dailykos. com/ story/2015/12/25/1463505/-Not-Just-Exxon-The-Entire-Oil-and-Gas-Industry-Knew-The-Truth-And-Decided-To-Lie-About-It.

"Not So Fast, Natural Gas! Why Accelerating Risky Drilling Threatens America's Water," Food & Water Watch Fact Sheet, July 2010, http:// www.foodandwaterwatch.org/sites/default/files/not_so_fast_natural_ gas_fs_july_2010.pdf.

Notice of Public Hearing regarding Docket #CO-14-020, the application of Hilcorp Alaska LLC (Hilcorp) for an exception to the spacing requirements of 20 AAC 25.055(a)(2) and 20 AAC 25.055(a)(4) to test and produce the Paxton No. 6 well within the Ninilchik Unit Undefined Tyonek and Beluga Gas Pools, comments to be received no later than July 28, 2014, State of Alaska, Alaska Oil and Gas Conservation Commission.

Oil and Gas Production in North Slope and Cook Inlet Over the Years, Alaska Department of Natural Resources, Division of Oil and Gas, http://dog.dnr.alaska.gov/Recruitment/AlaskaOilGas.htm.

Order Granting Long-Term Multi-Contract Authorization To Export Liquefied Natural Gas by Vessel from the Proposed Alaska LNG Project in the Nikiski Area of the Kenai Peninsula, Alaska, to Free Trade Agreement Nations; United States of America Department of Energy Office of Fossil Energy, DOE/FE Order No. 3554, Nov. 21, 2014.

"Pipeline building a choreography of coordinated steps," by Larry Persily, Kenai Peninsula Borough Mayor's Office, July 7, 2015.

"PCAST (President's Council of Advisors on Science and Technology) Releases New Climate Report," by Rick Weiss, March 22, 2013, https://www.whitehouse.gov/blog/2013/03/22/ pcast-releases-new-climate-report.

Right-of-way Lease For the Kenai Kachemak Pipeline, ADL 28162, Alaska Department of Natural Resources, Nov. 26, 2002.

South Ninilchik Unit Application, Findings and Decision of the Director, Division of Oil and Gas, Dec. 30, 2001.

"The negative effect of natural resources on development," by Melissa Dell, Alexander's Gas & Oil Connections, an Institute for Global Energy Research, Feb. 12, 2005, http://www.gasandoil.com/news/ features/fdaab7531289516b8792571d6ca7a115.

The Natural Gas Production Industry, U.S Environmental Protection Agency, https://www3.epa.gov/airquality/oilandgas/basic.html.

"Tokyo Gas sees Alaska in potential LNG supply mix," by Larry Persily, Kenai Peninsula Borough Mayor's Office, June 9, 2015.

Understanding Alaska's Revenue, http://www.alaskabudget.com/revenue.

"Vote No on 1 will fight repeal of oil tax reform," Straight Talk, published by the Alaska Oil and Gas Association, December 2013.

"Washington and the Oil Industry Know the Truth about Climate Change," by Dave Lindorff, Nation of Change, Jan. 19, 2014, http:// www.nationofchange.org/washington-and-oil-industry-know-truth- about-climate-change-1390149398.

INDEX

ACKNOWLEDGEMENTS

THE FIRST TIME I TOLD SOMEONE I WAS WRITING a book I was in the fourth grade. Who would have guessed it would take so long? Others' belief that it was more than a childish imagining bridged the years, and the personal accounts filling these pages brought the dream to life. May my retelling reflect the spirit in which these accounts were given. To everyone who participated, I am honored by the sharing of your experiences and thankful for what I learned from you. And to everyone who said, "You need to talk to . . . ," you've played an important part in getting these stories told. For permission to quote them at length, thanks to biologist Page Spencer, whose account of the aftermath of the *Exxon Valdez* spill in her book *White Silk and Black Tar* provides a vivid account of the enormity of size and cost of that tragedy, and to Alaskan Writer Laureate Nancy Lord, whose *Early Warming: Crisis and Response in the Climate-Changed North* is a never-more-urgent reminder of what's at stake. Activist Mike O'Meara's artwork strikes a chord with its depiction of the footprint oil and gas has made on Alaska.

To *Homer News* publisher Lori Evans: Thank you for inviting me into the world of writing way back when. To the *Homer News* team and especially to former co-worker Michael Armstrong, thank you for setting the bar so high. Author Marianne Schlegelmilch's commitment to writing and her success have been an inspiration. Adventurer Christina Whiting's writing is a constant reminder that the world is filled with beauty. Monthly get-togethers with my hometown buddies Gayle Forrest, Milli Martin and Randi Somers have offered breaks of good food and even better laughter. Untangling my attempts at knitting with the help of Anna Williams and the other women in the Driftwood Village stitch group, plus their frequent inquiries about the book's progress, smoothed the way through several writing challenges.

I owe a double debt of gratitude to Mark Wiltzius and Jeff Tachik at The Buzz in Ninilchik for the use of their wifi—and their excellent mochas that have fueled many a beach walk. Library Director Ann Dixon and her staff at the Homer Public Library provided a quiet place to interview, research and write. Librarian Victoria L. Steik and the volunteers at Ninilchik Community Library also offered a quiet corner for uninterrupted hours of research and writing, as did Library Director Jill E. Tierce and the clerks at Waldport Public Library in Waldport, Oregon, my winter home-away-from-home. And whether I was in Alaska or Oregon, partners Sue Post, Lee Post and Jenny Stroyeck of the Homer Bookstore unfailingly managed to fulfill my reference book needs.

Ken and Gail Lehman's oft-repeated question, "How's the book going?" was a gentle nudge to keep on keeping on until I could say, "It's written." Debi Rodriguez's innumerable expressions of support and Kathleen Bielawski's read-throughs, insights, suggestions and wicked humor helped keep the momentum going.

From the moment I approached her with the idea for this book, Hardscratch Press editor/publisher Jackie Pels has been an enthusiastic and constant pillar of support and expertise for which I am deeply grateful. She and her talented team of designer David Johnson, compositor Dickie Magidoff, proofreader Leah Pels and indexer Andrea Avni made a handsome and accessible book from a stack of manuscript pages.

The companionship of my sister, Risa Jackinsky, has proven an invaluable and steadfast anchor.

For all they've experienced, the trust of my daughters, Jennifer Long Stinson and Emily Long Aley, remains a gift I do not take lightly.

And throughout, thanks to my husband Sandy Mazen, who believes in this book, its message and me.

MCKIBBEN AUTUMN JACKINSKY'S Russian-Alutiiq great-great-great-grandparents were among the founders of Ninilchik village on the Cook Inlet side of the Kenai Peninsula. As a longtime Alaska journalist she has reported on the oil and gas industry from several perspectives. Now, with what author-activist Adam Briggle calls "an all too rare open-mindedness," she has interviewed families affected pro or con by the industry's presence in the area, as well as civic leaders, alternative energy advocates and others. In four unsparing chapters woven through *Too Close to Home?* she also tells her own family and personal story, on the way to a decision about oil and gas exploration on her inherited three-acre share of Jackinsky land. Most of the book was written in her Ninilchik cabin; otherwise she and her husband, Sandy Mazen, divide their time between Homer, Alaska, and Waldport, Oregon, for easier visiting with children and grandchildren.

Too Close to Home?
Living with "drill, baby" on Alaska's Kenai Peninsula

Project coordinator and editor: Jackie Pels
Book design and production: David R. Johnson
Typography and composition: Dickie Magidoff
Index by Andrea Avni, Vashon Island WordWrights
Proofreading: Leah H. Pels

Note: The only connection between author McKibben Jackinsky, whose first name is her mother's family name, and Bill McKibben, founder of 350.org and author of *Deep Economy* and other works, is their commitment to a healthy planet.

Printed and bound at McNaughton & Gunn, Saline, Michigan
Alkaline pH paper (Natural Offset)

Hardscratch Press
658 Francisco Court
Walnut Creek, CA 94598-2213
925/935-3422

[HARDSCRATCH]
www.hardscratchpress.com